THINKING BIG DATA IN GEOGRAPHY

THINKING
BIG DATA IN
GEOGRAPHY

NEW REGIMES,
NEW RESEARCH

Edited and with an introduction by
Jim Thatcher, Josef Eckert, and Andrew Shears

University of Nebraska Press | Lincoln and London

All rights reserved. Manufactured
in the United States of America ∞

Library of Congress
Cataloging-in-Publication Data
Names: Thatcher, Jim, 1980– editor.
Title: Thinking big data in geography: new
regimes, new research / edited and with an
introduction by Jim Thatcher, Josef Eckert,
and Andrew Shears.
Description: Lincoln: University of Nebraska
Press, [2018] | Includes bibliographical
references and index. |
Identifiers: LCCN 2017026971 (print)
LCCN 2017052109 (ebook)
ISBN 9781496205353 (epub)
ISBN 9781496205360 (mobi)
ISBN 9781496205377 (pdf)
ISBN 9780803278820 (cloth: alk. paper)
ISBN 9781496204981 (pbk.: alk. paper)
Subjects: LCSH: Geography—Data processing. |
Big data. | Geospatial data.
Classification: LCC G70.2 (ebook) |
LCC G70.2 .T45 2018 (print) |
DDC 910.285/57—dc23
LC record available at
https://lccn.loc.gov/2017026971

Set in Garamond Premier Pro
by Mikala R Kolander.

CONTENTS

ILLUSTRATIONS

TABLES

INTRODUCTION

Jim Thatcher, Andrew Shears, and Josef Eckert

This is a book about what, if any, "home field advantage" the discipline of geography might hold with "big data" given its history of dealing with large, heterogeneous sets of spatial information.[1] Contributing authors were asked what new avenues for knowledge and capital accumulation have been enabled and constrained by the purported data deluge.[2] In other words, what happens when "where" is recorded alongside who is doing what, when, and with whom?[3]

At the time the contributing authors were approached, in late 2014, the most exaggerated claims of the boosters of big data—those of "numbers speaking for themselves" and the "end of theory"—were already becoming the focus of criticism, morphing into the shibboleths by which those skeptical of big data could signal their belonging and launch their critiques.[4] Meanwhile studies of the geoweb and neogeography were calling attention to the ways in which user-generated data both come into the world and are complicit in its unfolding. Even as urban planners, politicians, marketers, national funding agencies, and the U.S. federal government embraced the generation, capture, and analysis of new forms of data as a primary tool by which to interpret the world, scholars were voicing caution regarding the uses of big data.[5] Scholars had called attention to issues with accuracy, heterogeneity of data and sources, surveillance, privacy, capital investment, and urban experience.[6]

Work in these and related areas has obviously continued.[7] But this book is a collection of pieces that stemmed from that original charge. On one hand a book is always a difficult format for the discussions and analyses of a rapidly evolving technological landscape. New applications, new formats of data, and even the legal terms by which researchers may access spatial data shift at a pace that far exceeds that of traditional forms of peer review and publication.[8] This technology-driven acceleration has led researchers to search for new publishing models and to adopt new terminology to better capture the nebulous, shifting terrain of their research. From the critical geographic information system (GIS) to the geoweb to critical data studies and beyond, we find fault with neither the continued search for new forms of discourse nor the drive for more accurate and precise terminology to describe the impacts of socio-technical advances.

However, books matter. As a discursive material object, this book matters because it represents the gathering of a variety of minds—from diverse fields and at disparate points in their careers—to discuss an overarching issue: what is big data and what does it mean to take it as both a means and object of research? As a collection of a dozen peer-reviewed chapters, plus introduction, by seventeen authors, this book offers multiple and sometimes conflicting answers to that question. Like data, these chapters ultimately only capture static moments as slices of thinking at specific time-spaces. Brought together they represent a deep, sustained engagement with the question at hand through a wide variety of important lenses. Whereas some chapters highlight the critical, epistemological limitations of big data (chapters 1, 2, 3, and 4), others espouse its very real potential to improve everyday understandings of climate change (chapter 10). Still others examine big data's impact on the cultural and political experience of urban landscapes (chapters 5, 7, 8, and 9). Our intention as editors, realized through this collection of sometimes discordant chapters, is to best represent the chaotic and ever-changing nature of contemporary big data studies within the discipline of geography.

For this reason we have eschewed formal definitions of big data and other terms in this introduction. As we have noted elsewhere, definitions for both will shift with the specific research focus of a piece.[9] Instead we allow each piece to stake its own claim to meaning. Here at the outset we instead present four overarching themes found coursing throughout this book that best reflect our own understandings of big data and its relations to geography: (1) the epistemologies of big data; (2) the shifting, complex nature of the "voluntary" production of data; (3) a dialectic of hope

Thatcher, Shears, and Eckert

and fear that runs through understandings of technology; and (4) the qualitative nature of purported quantified data. To address these themes the chapters of this book are organized into the following five sections: exploring the definitions of big data and what it means to study it, methods and praxis in big data research, empirical interventions, urban data, and talking across borders.

A short conclusion by Mark Graham connects many of the major themes, tying them together by exploring what an emerging critical study of big data might resemble. The remainder of this introductory chapter first explores the larger themes presented by this volume, then summarizes each chapter while highlighting their engagement with these questions.

Big Data as Epistemology

As a technical construct, big data is best understood as an ever-shifting target; as Jacobs puts it, big data is "data whose size forces us to look beyond the tried-and-true methods that are prevalent at that time."[10] Such a definition shows data to have always been big, encompassing the automatic tape array that first digitized the 1980 U.S. Census as well as the terabytes of data produced by the Large Hadron Collider today. However, somewhere along the "relentless march from kilo to tera and beyond," big data becomes an ideological orientation toward what constitutes both knowledge and its production.[11] This transformation is unsurprising and follows many of the same motivations and claims of neogeography and the geoweb itself.[12] As mentioned, a universal definition of big data is difficult to come by, both in this book and elsewhere. While different chapters highlight specific aspects of what constitutes big data, with many relying on some variation of the three-V trope of volume, velocity, and variety, an overarching theme is understanding big data as an epistemological stance.[13] In such a view big data is not only the physical infrastructure, the algorithms, and the ontologies that necessarily go into any sufficiently large ordering of data but also a stance that, as O'Sullivan puts it (chapter 2), "given sufficient data, the world can be known (if not completely, then well enough for any particular purpose)."

Despite big data's self-insistence on a sui generis origin story, viewing big data as an epistemology makes clear that its roots lie in older processes and concepts. For example, Bell, Hey, and Szalay have argued that, ever since the wide-scale adoption of the scientific process as a theoretical and experimental basis for knowledge production in the seventeenth century, scientists have consistently sought to create and analyze ever-larger data sets as a means of directly improving understandings of the physical universe.[14] Similarly Linnet Taylor has illustrated the stark parallels between the excitement around big data today and similar enthusiasm that surrounded the rise of the field of statistics in the eighteenth century.[15] Other authors have noted the roots of big data within social physics, geodemographics, and geomatics.[16] Considered in the context of larger processes of capitalist modernity, the epistemological commitments of big data clearly follow a distinct genealogy that runs back several centuries.

Reducing and representing the world with numbers only works in so far as said world may be remade in the image of those numbers.[17] Running through this book is a critical questioning of how those numbers are formed. Data are never raw; they are always cooked and must be "imagined as data to exist and function as such."[18] As such, the claims of big data are ideological ones that come with certain sets of epistemological commitments and beliefs. The chapters that follow deepen and extend understandings of what it means to live in a world infused with data, algorithms, and code.

Participation: Voluntary, Conscripted, or Something Else?

Both the digital divide and the uneven surfaces of data across time and space suggest a larger question: is participation in the generation of big data and other new regimes of data accumulation voluntary, conscripted, or something else entirely?[19] To answer requires more nuance than this question suggests, because the methods used to encourage participation are wide-ranging.

Many technologies that contribute to the generation of big data operate under a model in which users legally consent simply by using the technology itself, as governed by the product's terms of service (ToS) or end user license agreement (EULA). Despite empirical evidence that these often lengthy and legally framed documents are not read, they remain a key site at which individuals become dispossessed from the data they create.[20] One common example of this moment is found in the iTunes terms and conditions statement, upon which agreement is required by Apple iPhone owners before they can access the iTunes interface necessary for the device's online use—at least, without hacking or "jailbreaking" the device, a process requiring additional knowledge and skills to complete. The latest form of the iTunes terms and conditions statement comprises some 20,658 words; by other measures it is nearly six times the length of the Magna Carta and consists of nearly five times as many words as the U.S. Constitution. Consent to this document, and to participating in the big data project, becomes the price of entry for most persons.

Even beyond basic use of mobile and digital technologies, many activities that were previously beyond the purview of data collection have become sites for the production of not only "big" but also "small" forms of data (see chapter 3 for an exploration of the differences). Commercial outlets, such as supermarkets and pharmacies, increasingly have mandatory loyalty card memberships, which track and correlate purchasing habits, while many public spaces have become sites for algorithmically monitored video recording.[21] In such systems it becomes questionable as to whether individuals can opt out of data collection, with their options reduced to boycotting whole swaths of everyday life or to participating in regimes of data collection.[22] With these circumstances in mind it is worth asking again to what degree any given piece of data was knowingly and willingly contributed. Many of the chapters in this volume address this question in some way, from David Retchless's look at how informed, volunteered visualizations may influence climate science to Matthew Kelley's look at the new forms the digital divide has taken in recent years. Questions of hidden bias and how to address it appear in many chapters, such as Fekete's and Ricker's. Together these chapters illustrate how the possibility of avoiding the seemingly ever-expanding reach of big data, small data, and other new mapping technologies has become increasingly tenuous as issues of consent and participation blur.

Hope and Fear in Data

The ambiguity in the question of consent signals the more crucial, wider-reaching consequences of big and spatial data projects. Framing technology's role in the world as a double-edged sword runs through writing on the topic. Technology enables and constrains actions in the world: it destabilizes labor relations while opening new sites for surplus capital absorption, and in this way technology is "both friend and enemy."[23] Kingsbury and Jones suggest that the Frankfurt school's view on technology can be read as a broad dialectic between hope for technology's role as liberator and fear of its domination of everyday life, and we wish to extend this heuristic for this book.[24] We do not make a claim upon the theoretical orientation of any individual author nor reduce their claims to some structural box in which they can be placed. Rather it is to suggest that *if* we are to take seriously big data as a specific instantiation of technology in the world, *then* it is only natural to see outcomes that leave us both hopeful and fearful.

As such, the chapters in this book engage these topics from a variety of perspectives, probing the ways in which data functions in the world, while avoiding falling into hard technological determinism. Lingel (chapter 8) describes "watching big data happen" and how—in moments of hubris—alternative voices and visions can be swept away by powerful normative forces. Retchless (chapter 10) explores the distinct potential for these self-same technologies to improve the public's understanding of climate change. Our point here is not to offer an either/or proposition, wherein big and spatial data projects will either change the world for the better or enroll us within oppressive regimes of quantification; rather we seek to offer these and other possibilities as a *not only . . . but also* dynamic that, while proving more difficult to resolve conceptually, offers space for a wide diversity of approaches.[25]

As Graham illustrates in the final chapter, while the specific terms may change, the underlying processes of neither big data nor new web-mapping technologies are going to disappear. Unless we want to ignore the project in its totality, we must ask important critical questions about this new paradigm, not only who is being left out and who is being exploited but also how new sources of data "help us to answer the big questions that we need to ask." It is impossible, or at least irresponsible, to be blind to for-profit motivations behind many new spatial data and big data firms; however, it is similarly irresponsible to not consider, propose, and practice alternatives that take up the banner of justice, equity, and social good as their core objective. That innate tension runs, by design, through the chapters of this book.

Seeing the Qualitative

Further epistemological tensions within big and spatial data arise from the nature of data and its analysis. Contemporary big data practices have often been undergirded by a resurgent pseudopositivism that accepts quantification uncritically.[26] With respect to social media and geodemographic data, big data comes to represent the individual who created it, reducing the complexity of human experience to a limited set of purportedly quantitative variables.[27] As illustrated above, this desire to reduce the world to numbers has its roots in much older tendencies toward statistical analysis within capitalist modernity.[28] However, that is not to suggest quantitative analysis and methodologies have no place in knowledge production. Through this book we instead want to suggest a need to see the qualitative nature within quantitative data. Even where the rigor of statistical analysis has produced empirical, robust results working with new, large, heterogeneous data sets, we want to suggest a moment of reflection on the construction of code, data, and algorithms.

The chapters of this text offer different insights into how to question the qualitative within the quantitative. Ricker argues in chapter 4 that the analysis and visualizations of big data are always inevitably "influenced by . . . epistemolog[ies]" of the researchers involved. By seeing the qualitative within the quantitative, Ricker demonstrates how the rigor of qualitative methodologies can strengthen datacentric analysis. Chapters like Fekete's exploration of Foursquare check-ins in Kansas City (chapter 7) and Jung and Shin's work on Washington State election tweets (chapter 5) attempt to directly bridge the supposed gap between quantitative and qualitative, exploring the limits at which grounded theory, qualitative methods, and quantitative big data meet. Ultimately there is no single answer here, or elsewhere, as to the exact limits of qualitative and quantitative methods. In this book the authors grapple with the limits of big data and the importance of understanding where the qualitative, affective moments of human life are constrained by moments of classification of digital information.

Organization of the Volume

Chapters in this volume have been organized into five sections, the divisions of which are based loosely upon how the author(s) approach big data and geography and how the chapters engage with the themes extrapolated above.

What Is Big Data and What Does It Mean to Study It?

In chapter 1 Kitchin and Lauriault explore a new vision for critical data studies (CDS) in geography and how such an epistemology would provide significant insight into a myriad of significant questions critical researchers should be asking about the provenance of big data. Building from the work of Dalton and Thatcher, Kitchin and Lauriault forward the data assemblage—an agglomeration of factors that contribute to data's creation, circulation, and application, including technological, social, economic, and political contexts crucial to framing the data at hand—as a unit for critical analysis.[29] The authors draw on Foucault and on Hacking as theoretical guideposts for unpacking these assemblages as a starting point for CDS, providing illustrations of how such assemblages have impacts far greater than the sum of their parts.

Recognizing the wide-scale adoption of big data as an important data source for computational studies within the social sciences, O'Sullivan (chapter 2) calls for an adjustment in the epistemology used to understand these data—from examining the novelty of the data themselves to a better use of computational frameworks when leveraging such data to explain the world. Citing the ascendancy of certain big data methodologies that value data volume over all else, the author demonstrates how a specific form of computational social science has accompanied this rise, one based on identification of variables and the establishment of math-

xviii

Thatcher, Shears, and Eckert

ematical relationships between them. Demonstrating the inadequacy of such approaches, O'Sullivan explores approaches that better recognize and represent processes. He concludes by arguing for the geographic application of approaches taken from complexity science, a field that has been largely ignored in geography since the 1980s and 1990s.

Methods and Praxis in Big Data Research

Citing several concerns with the big data paradigm, chapter 3 authors Sieber and Tenney forward a counterargument to the notion that bigger big data is always better by exploring the problematic binary used to differentiate big data from "small data." While remaining "agnostic about the value of big data and data-science methodologies," the authors urge caution about couching all data studies within the buzzy and evolving big data epistemology. Moving through various potential definitions of big and small, the authors explore how the very constitution of data as an object of research shifts across scales. To Sieber and Tenney some of the shortcomings of a perspective prioritizing the size of big data can be solved by continuing to acknowledge the legitimacy of small data and small data–driven studies, even within the big data paradigm.

Another proposal for refining the big data paradigm comes from the author of chapter 4. In her chapter, Ricker convincingly argues that big data, and especially spatial data, is mostly qualitative in nature. Despite the tendency of many big data researchers and practitioners, driven by the intimidating size of such data sets, to focus exclusively on quantitative readings and analyses, Ricker suggests that aspects of qualitative methodologies, including acknowledgment of subjective approaches to issues of reflexivity and positionality, can provide a rigor largely missing from current big data projects.

Empirical Interventions

Recognizing the limited focus of many spatial data studies in terms of the acquisition of massive data sets for quantitative analysis, chapter 5 authors Jung and Shin argue that a hybrid qualitative-quantitative approach to such work allows for researchers to minimize inherent issues with such data sets by providing a social and linguistic context for the data points. Jung and Shin then apply their proposed mixed-method approach, which combines quantitative techniques, including geographic/temporal visualization and spatial analysis, with a qualitative ethnographic reading of data powered by grounded theory, to a collection of tweets from the Seattle area during debates on legalization of marijuana and same-sex marriage. Through this effort the authors demonstrate that some of the more commonly cited limitations of spatial data are not absolute.

Acknowledging the wide-scale privacy and consent concerns inherent to spatial big data and recognizing that users theoretically volunteering this information may have no real idea of how often those data are accessed, chapter 6 authors Weidemann, Swift, and Kemp introduce a web application that allows users to assess privacy concerns applicable to their online social media activity. The resulting application, Geosocial Footprint, offers a tool that opens the door to alternative approaches to empowering end users to confront their data in an online environment.

The availability of volunteered geographic information, particularly in the form of geotagged public social media quotes, has been a particularly fruitful path toward publication for geographers. However, use of such data comes with a number of caveats and limitations that researchers are still struggling to fully explicate. In chapter 7 Fekete reports results from a case study of data from a location-based social media network (Foursquare) found in a localized area (Kansas City) as a means of demonstrating selection bias within available social media data. In this study she compares patterns visible from check-in data to measures of neighborhood demographics as tracked by the U.S. Census, finding that the demographic characteristics of the Foursquare user base are vastly different from the known demographic measures of Kansas City neighborhoods. Fekete thus empirically demonstrates significant socioeconomic bias within the Foursquare data, showing that the app favors more affluent and whiter populations.

Urban Big Data: Urban-Centric and Uneven

In chapter 8—a short, autoethnographic piece examining the impact of big data on urban landscapes of sexuality—Lingel explores both metro-

Thatcher, Shears, and Eckert

normativity (via Halberstam) and the resulting implications for queer urban spaces brought forth by visible and material incarnations of big data. Using personal experience as a contextual framework and incorporating questions of privacy, disempowerment, and big data infrastructure, Lingel calls for an adjustment to ethical questions concerning new data regimes in order to incorporate the impact of these technologies on the urban landscape, not for a faceless majority but for those who actually work and interact within that place.

Using a long-established literature regarding issues of the so-called digital divide (inequality of access to the Internet and related technologies), chapter 9 author Kelley writes an illustrative piece examining the impacts of such inequality on the urban landscape in an age in which mobile and wearable technologies have become commonplace. Kelley demonstrates how, as these technologies increasingly constitute and mediate the urban experience—governing everything from the use of public transit to equal access to nominally public spaces—the digital divide has not disappeared but rather has become more nebulous and difficult to reconcile. Kelley suggests a research orientation that recognizes the increasing costs of living on the wrong side of this divide, one that understands the issue not as simply access to a set of technologies but also as the education and cultural norms that relate to their use. Kelley concludes by noting that the integration of mobile geospatial technologies and the urban landscape has occurred only within the most recent decades and is likely to change many times over the coming years before a "technological equilibrium" can be achieved. We must, as researchers and as a public, work to ensure such an equilibrium is just and equitable.

Talking across Borders

Seeking to address the popular intellectual disconnect between climate change and its anthropogenic causes, Retchless proposes in chapter 10 a novel use of new spatial data visualization technologies as means of exploring a global phenomenon at more immediate scales. The author explores the many barriers that climate scientists face in communicating the consequences of continued human-forced change, including its scale and complexity, predictive uncertainty, and the difficulty of experiential observation attributed to the at times seemingly contradictory conditions at local and global scales. To combat these concerns Retchless proposes enhanced citizen engagement through two approaches—utilizing citizen sensors and personal visualizations—and evaluates how this engagement can further enhance climate change literacy among the citizenry.

With the advent of participatory, technologically mediated approaches to the answering of large-scale geographic questions, a large group of researchers and practitioners have begun to espouse so-called Web 2.0 approaches to the humanitarian work. In chapter 11, despite what Burns terms the "inherent spatialities" of digital humanitarian work, he critiques the paucity of attention that has been paid to this topic by researchers within geography. Burns argues that this occurs despite the attempts of those working in the digital humanities to crowdsource the (often geospatial) data needed for humanitarian purposes by engaging volunteers to gather, produce, and process it. Through a literature review of contemporary digital humanitarian work and vivid illustrations provided by ethnographic interviews, Burns demonstrates that digital humanitarianism is intrinsically linked to and can be best understood as a specific manifestation of new regimes of spatial data generation, acquisition, analysis, and visualization.

In lieu of an editorial conclusion Graham offers a series of pointed interjections for big data researchers to ponder as they conclude the volume. Graham takes a step back from the immediacy of research to ask where the field of study stands moving forward. Just as Kitchin and Lauriault began with an extension of Dalton and Thatcher's work on critical data studies, Graham outlines his own extension of that work, one that recognizes both that current mixed-method approaches to big data have rung hollow and that geography, as a discipline, is always constantly fighting its own insular tendencies. He urges geographers to apply a more critical edge to their studies, noting that "platforms and mediators that we rely on . . . do not necessarily have issues of justice, equality, human rights, and peace" as priorities. In order to address these topics we must look within and without; we must recognize inherent issues of privacy and bias, seeing the qualitative in the quantitative. At the same time, we must not forget the physical materialities of digital data, both in terms of servers and electricity, as well as in terms of the hidden labor that goes into its creation and maintenance.

In order to avoid constantly reinventing the wheel, Graham, like many of the other authors in this volume, implores us to look to spatial work being done in a variety of disciplines. Here we would like to extend that examination not only to other disciplines currently but also, following O'Sullivan, to other times within our own discipline. To reiterate, this is part of why this book matters—it distills the thinking on these topics at a particular time and in a particular space. It cannot cover all there is to say about big data but instead hopes to open up a series of new collaborations and questions. The ideological and socioeconomic forces that constitute big data aren't going away, even if any given specific term for their study may disappear from the peer-reviewed corpus in the coming years. In this specific moment, before the new regimes of data creation, extraction, and analysis recede fully from conscious consideration and become yet another aspect of modern life, we call for a moment of reflection: a moment of critical inquiry into what it means to study big data as a geographer. Despite the recognized and repeated need to critique big data and its seemingly interminable quest to mediate everyday life, we agree with Thatcher et al. that the present reflects a particular moment of optimism for the forging of new alliances within and across disciplines.[30] Ultimately this book gathers a set of voices that, while divergent in perspective, are united in their drive to understand the crevasses and cracks of big data and to find those gaps and moments that leave space for interventions within a world increasingly mediated by geospatial technologies.

Notes

1. Farmer and Pozdnoukhov, "Building Streaming GIScience," 2.
2. Anderson, "End of Theory"; *Economist*, "Data Deluge"; Baraniuk, "More Is Less"; Kitchin and Dodge, *Code/Space*.
3. Feenberg, *Critical Theory of Technology*; Kitchin, *Data Revolution*.

4. C. Anderson, "End of Theory"; M. Graham, "Big Data"; Kitchin, *Data Revolution*; Kitchin and Dodge, *Code/Space*.

5. Torrens, "Geography and Computational Social Science"; Morozov, *To Save Everything*; Nickerson and Rogers, "Political Campaigns and Big Data"; LaValle et al., "Big Data, Analytics"; Mayer-Schönberger and Cukier, *Big Data: Revolution*; National Science Foundation, "Critical Techniques, Technologies and Methodologies"; Research Councils UK, "Big Data"; Executive Office of the President, *Big Data*.

6. Crawford, "Hidden Biases in Big Data"; Goodchild, "Quality of Big (Geo)Data"; Kitchin, *Data Revolution*; Stephens, "Gender and the GeoWeb"; Crampton, *Mapping*; Crampton et al., "Beyond the Geotag"; Elwood and Leszczynski, "Privacy, Reconsidered"; Bettini and Riboni, "Privacy Protection"; Wilson, "Location-Based Services"; Thatcher, "Avoiding the Ghetto"; Zheng and Hsieh, "U-Air."

7. See, for example, Thatcher, O'Sullivan, and Mahmoudi, "Data Colonialism through Accumulation"; Thakuriah, Tilahuan, and Zellner, "Big Data and Urban Informatics"; Leszczynski, "Spatial Big Data"; Crampton et al., "Beyond the Geotag"; and Zhong et al., "Variability in Regularity," among many others.

8. Thatcher, "Big Data, Big Questions."

9. Thatcher, "Big Data, Big Questions"; Dalton and Thatcher, "Critical Data Studies."

10. Jacobs, "Pathologies of Big Data."

11. Doctorow, as quoted in Thatcher, O'Sullivan, and Mahmoudi, "Data Colonialism through Accumulation."

12. Leszczynski, "On the Neo in Neogeography."

13. Laney, 3D *Data Management*; boyd and Crawford, "Critical Questions for Big Data."

14. Bell, Hey, and Szalay, "Beyond the Data Deluge."

15. Dalton, Taylor, and Thatcher, "Critical Data Studies."

16. Barnes and Wilson, "Big Data, Social Physics"; Dalton and Thatcher, "Critical Data Studies"; Karimi, *Big Data: Techniques and Technologies*.

17. Porter, *Rise of Statistical Thinking*.

18. Gitelman and Jackson, "Introduction," 3.

19. Kelley, "Semantic Production of Space" and in chapter 9 of this volume; Dalton, Taylor, and Thatcher, "Critical Data Studies."

20. J. Lin et al., "Expectation and Purpose"; Thatcher, O'Sullivan, and Mahmoudi, "Data Colonialism through Accumulation."

21. Kitchin, "Big Data, New Epistemologies."

22. Lanier, *You Are Not a Gadget*.

23. Feenberg, *Critical Theory of Technology*; Harvey, *Condition of Postmodernity*; Harvey, *Enigma of Capital*; Postman, *Technopoly*, quoted in Naughton, "Technology Is a Double-Edged Sword."
24. Kingsbury and Jones, "Walter Benjamin's Dionysian Adventures."
25. Barnes, "'Not Only . . . But Also.'"
26. Wyly, "New Quantitative Revolution," 30.
27. Thatcher, O'Sullivan, and Mahmoudi, "Data Colonialism through Accumulation."
28. Scott, *Seeing Like a State*; Foucault, *Birth of Biopolitics*.
29. Dalton and Thatcher, "Critical Data Studies."
30. Thatcher, O'Sullivan, and Mahmoudi, "Data Colonialism through Accumulation."

PART 1

What Is Big Data and What
Does It Mean to Study It?

1 Toward Critical Data Studies

Charting and Unpacking Data Assemblages and Their Work

Rob Kitchin and Tracey P. Lauriault

A Critical Approach to Data

Societies have collected, stored, and analyzed data for a couple of millennia as a means to record and manage their activities. For example, the ancient Egyptians collected administrative records of land deeds, field sizes, and livestock for taxation purposes, the Domesday Book in 1086 captured demographic data, double-entry bookkeeping was used by bankers and insurers in the fourteenth century, and the first national registry was undertaken in Sweden in the seventeenth century.[1] It was not until the seventeenth century, however, that the term "data" was used for the first time in the English language, thanks to the growth of science, the development of statistics, and the shift from knowledge built from theology, exhortation, and sentiment to facts, evidence, and the testing of theory through experiment.[2] Over time the importance of data has grown, becoming central to how knowledge is produced, business conducted, and governance enacted. Data provide the key inputs to systems that individuals, institutions, businesses, and the sciences employ in order to understand, explain, manage, regulate, and predict the world we live in and are used to create new innovations, products, and policies.

The volume, variety, and use of data have grown enormously since the seventeenth century, and there has long been the creation and maintenance of very large data sets, such as censuses or government administrative and natural resource databases. Such databases, however, have typically been generated every few years or are sampled. In contrast, over the past fifty years we have begun to enter the era of big data, with such characteristics as being

- huge in *volume*, consisting of terabytes or petabytes of data;
- high in *velocity*, being created in or near real time;
- diverse in *variety*, being structured and unstructured in nature;
- *exhaustive* in scope, striving to capture entire populations or systems (n = all);
- fine-grained in *resolution* and uniquely *indexical* in identification;
- *relational* in nature, containing common fields that enable the conjoining of different data sets; and
- *flexible*, holding the traits of *extensionality* (new fields can easily be added) and *scalability* (data sets can expand in size rapidly).[3]

While there are varying estimates, depending on the methodology used, as to the growth of data production caused in the main by the production of big data, in addition to a steep growth in small data such as personal video, photo, and audio files (all of which consume large amounts of data storage), it is clear that there has been a recent step change in the volume of data generated, especially since the start of the new millennium.[4] Gantz and Reinsel have estimated that data volumes had grown by a factor of nine in the preceding five years, and Manyika et al. have projected a 40 percent rise in data generated globally per year.[5] In 2013 EU Commissioner for the Digital Agenda Neelie Kroes reported that 1.7 million billion bytes of data per minute were generated globally.[6] Such rises and projections for further increases are due to the continuous and exhaustive, rather than sampled, production of born digital data, in combination with the nature of some of those data (e.g., image and video files) and the increased ability to store and share such data at marginal cost. For example, in 2012 Facebook reported that it was processing 2.5 billion pieces of content (links, comments, etc.), 2.7 billion "Like" actions, and 300 million photo uploads *per day*, and Walmart was generating more than 2.5 petabytes (2^{50} bytes) of data relating to more than 1 million customer transactions *every hour*.[7]

These massive volumes of data are being produced by a diverse set of information and communication technologies that increasingly mediate and augment our everyday lives, for example, digital CCTV, retail checkouts, smartphones, online transactions and interactions, sensors and scanners, and social and locative media. As well as being produced by government agencies, vast quantities of detailed data are now being generated by mobile phone operators, app developers, Internet companies, financial institutions, retail chains, and surveillance and security firms, and data are being routinely traded to and between data brokers as an increasingly important commodity. More and more analog data held in archives and repositories are being digitized and linked together and made available through new data infrastructures, and vast swaths of government-produced and held data are being made openly accessible as the open data movement gains traction.[8]

This step change in data production has prompted critical reflection on the nature of data and how they are employed. As the concept of data developed, data largely came to be understood as being pre-analytical and prefactual—that which exists prior to interpretation and argument or the raw material from which information and knowledge are built. From this perspective data are understood as being representative, capturing the world as numbers, characters, symbols, images, sounds, electromagnetic waves, bits, and so on, and holding the precepts of being abstract, discrete, aggregative (they can be added together), nonvariant, and meaningful independent of format, medium, language, producer, and context (i.e., data hold their meaning whether stored as analog or digital, viewed on paper or screen, or expressed in different languages).[9] Data are viewed as being benign, neutral, objective, and nonideological in essence, reflecting the world as it is subject to technical constraints; they do not hold any inherent meaning and can be taken at face value.[10] Indeed the terms commonly used to detail how data are handled suggest benign technical processes: "collected," "entered," "compiled," "stored," "processed," and "mined."[11] In other words it is only the uses of data that are political, not the data themselves.

This understanding of data has been challenged in recent years. Contrary to the notion that data is pre-analytic and prefactual is the argument that data are constitutive of the ideas, techniques, technologies, people, systems, and contexts that conceive, produce, process, manage, and analyze them.[12] In other words, how data are conceived, measured, and employed actively frames their nature. Data do not pre-exist their generation; they do not arise from nowhere, and their generation is not inevitable: protocols, organizational processes, measurement scales, categories, and standards are designed, negotiated, and debated, and there is a certain messiness to data generation. As Gitelman and Jackson put it, "raw data is an oxymoron"; "data are always already 'cooked.'"[13] Data then are situated, contingent, relational, and framed and are used contextually to try and achieve certain aims and goals.

Databases and repositories are also not simply a neutral, technical means of assembling and sharing data but are bundles of contingent and relational processes that do work in the world.[14] They are complex socio-technical systems that are embedded within a larger institutional landscape of researchers, institutions, and corporations and are subject to socio-technical regimes "grounded in . . . engineering and industrial practices, technological artifacts, political programs, and institutional ideologies which act together to govern technological development."[15] Databases and repositories are expressions of knowledge/power, shaping what questions can be asked, how they are asked, how they are answered, how the answers are deployed, and who can ask them.[16]

Beyond this philosophical rethinking of data, scholars have begun to make sense of data ethically, politically and economically, spatially and temporally, and technically.[17] Data can concern all aspects of everyday life, including sensitive issues, and be used in all kinds of ways, including to exploit, discriminate against, and persecute people. There are then a series of live moral and ethical questions concerning how data are produced, shared, traded, and protected; how data should be governed by rules, principles, policies, licenses, and laws; and under what circumstances and to what ends data can be employed. There are no simple answers to such questions, but the rise of more widespread and invasive data generation and more sophisticated means of data analysis creates an imperative for public debate and action. In addition data are framed by political concerns as to how they are normatively conceived and contested as public

Kitchin and Lauriault

and private goods. The open data and open government movements, for example, cast data as a public commons that should be freely accessible. Business, in contrast, views data as a valuable commodity that, on the one hand, needs to be protected through intellectual property regimes (copyright, patents, ownership rights) and, on the other, should be exploitable for capital gain. Indeed data often constitute an economic resource: for government they are sold under cost-recovery regimes and for business they are tradable commodities to which additional value can be added and extracted (e.g., derived data, analysis, knowledge). In the present era data are a key component of the emerging knowledge economy enhancing productivity, competitiveness, efficiencies, sustainability, and capital accumulation. The ethics, politics, and economics of data develop and mutate across space and time with changing regimes, technologies, and priorities. From a technical perspective, there has been a focus on how to handle, store, and analyze huge torrents of data, with the development of data mining and data analytics techniques dependent on machine learning, and there have been concerns with respect to data quality, validity, reliability, authenticity, usability, and lineage.

In sum we are starting to witness the development of what Dalton and Thatcher call critical data studies—research and thinking that apply critical social theory to data to explore the ways in which they are never simply neutral, objective, independent, raw representations of the world but are situated, contingent, relational, contextual, and do active work in the world.[18] In their analysis Dalton and Thatcher set out seven provocations needed to provide a comprehensive critique of the new regimes of data:

- situating data regimes in time and space;
- exposing data as inherently political and identifying whose interests they serve;
- unpacking the complex, nondeterministic relationship between data and society;
- illustrating the ways in which data are never raw;
- exposing the fallacies that data can speak for themselves and that big data will replace small data;

- exploring how new data regimes can be used in socially progressive ways; and
- examining how academia engages with new data regimes and the opportunities of such engagement.

We agree with the need for all of these provocations. In a short presentation at a meeting of the Association of American Geographers one of us set out a vision for what critical data studies might look like: unpacking the complex assemblages that produce, circulate, share/sell, and utilize data in diverse ways; charting the diverse work they do and their consequences for how the world is known, governed, and lived in; and surveying the wider landscape of data assemblages and how they interact to form intersecting data products, services, and markets and shape policy and regulation. It is to this endeavor that we now turn.

Charting and Unpacking Data Assemblages

Kitchin defines a data assemblage as a complex socio-technical system that is composed of many apparatuses and elements that are thoroughly entwined and whose central concern is the production, management, analysis, and translation of data and derived information products for commercial, governmental, administrative, bureaucratic, or other purposes (see table 1-1).[19] A data assemblage consists of more than the data system or infrastructure itself, such as a big data system, an open data repository, or a data archive, to include all of the technological, political, social, and economic apparatuses that frame their nature, operation, and work. The apparatuses and elements detailed in table 1-1 interact with and shape each other through a contingent and complex web of multifaceted relations. And just as data are a product of the assemblage, the assemblage is structured and managed to produce those data.[20] Data and their assemblage are thus mutually constituted, bound together in a set of contingent, relational, and contextual discursive and material practices and relations. For example, the data assemblage of a census consists of a large amalgam of apparatuses and elements that shape how the census is formulated, administered, processed, and communicated and how its findings are employed. A census is underpinned by a realist system of thought; it has a diverse set of accompanying forms of supporting documentation; its questions are negotiated by many stakeholders; its costs are a source of contention; its administering and reporting are shaped by legal frameworks and regulations; it is delivered through a diverse set of practices, undertaken by many workers, using a range of materials and infrastructures; and its data feed into all kinds of uses and secondary markets. Data assemblages evolve and mutate as new ideas and knowledges emerge, technologies are invented, organizations change, business models are created, the political economy changes, regulations and laws are introduced and repealed, skill sets develop, debates take place, and markets grow or shrink. And while data sets once generated within an assemblage may appear fixed and immutable (e.g., a compiled census), they are open to correction and revision, reworking through disaggregation and reaggregation into new classes or statistical geographies, parsing into other data systems, data derived and produced from them, and alternative interpretations and insights drawn from them. Data assemblages and their data are thus always in a state of becoming.

This notion of a data assemblage is similar to Foucault's concept of the *dispositif*, which refers to a "thoroughly heterogeneous ensemble consisting of discourses, institutions, architectural forms, regulatory decisions, laws, administrative measures, scientific statements, philosophical, moral[,] and philanthropic propositions" that enhance and maintain the exercise of power within society.[21] The dispositif of a data infrastructure produces what Foucault terms "power/knowledge," that is, knowledge that fulfills a strategic function: "the apparatus is thus always inscribed in a play of power, but it is also always linked to certain coordinates of knowledge which issue from it but, to an equal degree, condition it. This is what the apparatus consists in: strategies of relations of forces supporting, and supported by, types of knowledge."[22] In other words, data infrastructures are never neutral, essential, objective; their data are never raw but always cooked to some recipe by chefs embedded within institutions that have certain aspirations and goals and operate within wider frameworks.

TABLE 1-1. Apparatus and elements of a data assemblage

APPARATUS	ELEMENTS
Systems of thought	Modes of thinking, philosophies, theories, models, ideologies, rationalities, etc.
Forms of knowledge	Research texts, manuals, magazines, websites, experience, word of mouth, chat forums, etc.
Finance	Business models, investment, venture capital, grants, philanthropy, profit, etc.
Political economy	Policy, tax regimes, incentive instruments, public and political opinion, etc.
Governmentalities and legalities	Data standards, file formats, system requirements, protocols, regulations, laws, licensing, intellectual property regimes, ethical considerations, etc.
Materialities and infrastructures	Paper/pens, computers, digital devices, sensors, scanners, databases, networks, servers, buildings, etc.
Practices	Techniques, ways of doing, learned behaviors, scientific conventions, etc.
Organizations and institutions	Archives, corporations, consultants, manufacturers, retailers, government agencies, universities, conferences, clubs and societies, committees and boards, communities of practice, etc.
Subjectivities and communities	Data producers, experts, curators, managers, analysts, scientists, politicians, users, citizens, etc.
Places	Labs, offices, field sites, data centers, server farms, business parks, etc., and their agglomerations
Marketplace	For data, its derivatives (e.g., text, tables, graphs, maps), analysts, analytic software, interpretations, etc.

This cooking of data is revealed through the work of Ian Hacking, who drew inspiration from Foucault's thinking on the production of knowledge.[23] Hacking posits that within a data assemblage there are two interrelated processes at work that produce and legitimate its data and associated apparatuses and elements, shaping how its data do work in the world, that in turn influence future iterations of data and the constitution of the assemblage. In both cases he posits that a dynamic nominalism is at work, wherein there is an interaction between data and what they represent, leading to mutual changes.

The first of these processes is what Hacking terms the "looping effect."[24] The looping effect concerns how data are classified and organized, how a data ontology comes into existence, and how it can reshape that which has been classified. The loop (fig. 1-1) has five stages:

1. classification, wherein things that are regarded as having shared characteristics are grouped together or, in cases of deviance, forced into groupings;
2. objects of focus (e.g., people, spaces, fashions, diseases, etc.) wherein, in the case of people, individuals eventually start to identify with the class into which they are assigned or, in the case of nonhuman objects, people come to understand and act toward the objects according to their classification;
3. institutions, which institutionalize classifications and manage data infrastructures;
4. knowledge, which is used to formulate, reproduce, and tweak classifications; and
5. experts, being those within institutions who produce and exercise knowledge, implementing the classification.

Through this looping effect Hacking argues that a process of "making people up" occurs in data systems such as the census or the assessing of mental health, wherein the systems of classification work to reshape society in the image of a data ontology. Examples could include people defining themselves or being defined by mental health symptoms, as well as a system of mental health facilities being built and staffed by specialist professionals.

The second of the processes consists of what Hacking terms "engines of discoverability" that extend beyond simply methods. He discusses these methods using a medical lens, which Lauriault has modified to incorporate the making up of spaces as well as people.[25] Hacking posits that there are a number of such engines, the last three of which are derived engines that are

a. counting the volumes of different phenomena;
b. quantifying: turning counts into measures, rates, and classifications;
c. creating norms: establishing what might or should be expected;
d. correlation: determining relationships between measures;
e. taking action: employing knowledge to tackle and treat issues;
f. scientification: establishing and adopting scientific knowledge;
g. normalization: seeking to fashion the world to fit norms (e.g., encouraging diets to meet expected body mass indices);
h. bureaucratization: putting in place institutions and procedures to administer the production of expectations and to undertake action; and
i. resistance to forms of knowledge, norms, and bureaucracy by those who are affected in negative ways (e.g., homosexual and disabled people's resistance to medicalized models that class, position, and treat them in particular ways) or those forwarding alternative systems, interpretations, and visions.[26]

Together these engines undertake the work of a data assemblage at the same time as it legitimates and reproduces such work and the assemblage itself. For example, a census counts a population and aspects of people's lives, turns that information into measures, establishes baseline rates, assesses relationships between factors, and is transformed into knowledge, which leads to practices of normalization and is enacted by dedicated and related bureaucracy. Each stage reinforces the previous, and collectively they justify the work it does. The knowledge produced and indeed the whole assemblage can be resisted, as with the census boycotts in Germany in the 1980s or with campaigns to ensure that Irish ethnicity is not undercounted in the UK, that "New Zealander" is accepted as an ethnicity in New Zealand (instead of "New Zealand European"), and that women's unpaid work is accounted for, or the knowledge produced can be transgressed, as in the case of those who report their religion as Jedi.[27] It can indeed even be canceled, as in the 2011 long-form census of Canada.

Data assemblages form part of a wider data landscape composed of many interrelated and interacting data assemblages and systems. Within the public sector, for example, there are thousands of data systems (each one surrounded by a wider assemblage) that interact and work in concert to produce state services and forms of state control at the local, regional, and national levels. Often this data landscape extends to the pan-national and the global scale, through interregional and worldwide data sets, data-sharing arrangements and infrastructures, and the formulation of protocols, standards, and legal frameworks (e.g., Global Spatial Data Infrastructures, INSPIRE). Firms within industry likewise create and occupy a complex data landscape, selling, buying, and sharing data from millions of data systems, all part of wider socio-technical assemblages. For example, the data landscape of big data consists of hundreds of companies, ranging from small and local to large and global, that provide a range of complementary and competing services, such as cooked data, specialty compilers and aggregators, data analytics, segmentation tools, list management, interpretation and consulting, marketing, publishing, and research and development. We have barely begun to map out various data landscapes, their spatialities and temporalities, their complex political economy, and the work that they do in capturing, analyzing, and reshaping the world. It is to the latter we now turn.

Uncovering the Work of Data Assemblages

As noted in the previous section, data assemblages do work in the world. Data are being leveraged to aid the tasks of governing people and territories, managing organizations, producing capital, creating better places, improving health care, advancing science, and so on. This leveraging takes many forms, but the central tenet is that data, if analyzed and exploited

Looping Effect

- (i) classification
- (v) experts
- (ii) objects of focus
- (iv) knowledge
- (iii) institutions

Engines of discovery

(a) counting	(b) quantifying
(c) creating norms	(d) correlation
(e) taking action	(f) scientification

Derived engines

(g) normalization

(h) bureaucracy   (j) resistance

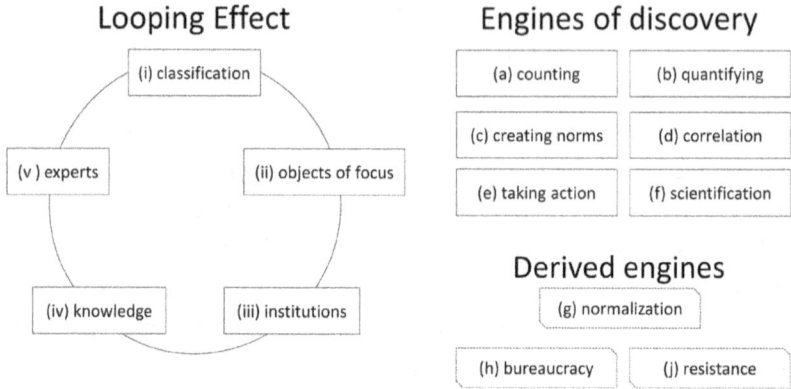

FIG. 1-1. The working of a data assemblage, following Hacking, "Philosophie et histoire des concepts scientifiques," and Laney, 3D *Data Management*. Created by R. Kitchin and T. Lauriault.

appropriately, produce information and knowledge that can be used to reshape operating procedures and organizational structure, identify new products, segment markets, reduce uncertainty and risk, and increase efficiency, productivity, competitiveness, and sustainability.[28] While much of the work to which data are put is beneficial to wider society, with data being used to improve quality of life and to tackle humanitarian and environmental issues, there is also a darker side to much data work. Here we want to consider the latter, highlighting four ways in which data are being employed to produce pernicious social and economic relations: dataveillance and the erosion of privacy, profiling and social sorting, anticipatory governance, and secondary uses and control creep. These practices are currently the subject of much debate, and there is an urgent need for critical studies that can inform the arguments being made.

As the revelations of WikiLeaks, Edward Snowden and other whistle blowers, the Maher Arar case, and other legal challenges with respect to erroneous record keeping and the mistreatment of individuals have demonstrated, from 9/11 onward there has been a step change in the extent and nature of state-led surveillance and securitization in many nations. Vast quantities of everyday communications (telephone calls, text messages, emails, social media), as well as general Internet use, are being routinely and systematically gathered by organizations such as the U.S. National Security Agency and analyzed for strategic intelligence.[29] All nation-states similarly gather large databases of information about citizens with respect to all aspects of their lives—income, tax, welfare, health, education, and so on. Likewise companies now routinely generate data with respect to all aspects of their business, including their customers and their patterns of consumption. Indeed given the mediating role of software in tasks such as working, traveling, consuming, communicating, and playing, it is increasingly difficult to take part in daily life without leaving a digital trace.[30] For example, the Dutch Data Protection Authority estimates that the average Dutch citizen is recorded in 250 to 500 databases, with some in up to 1,000 databases—a figure that is growing.[31] These databases not only include individuals' digital footprints (data they themselves leave behind) but also individuals' data shadows (information about them generated by others). Those to whom the data refer often have little control over the data generated, their form, extent, or how they are used.[32] Individually these databases provide limited views of people, but they gain power when combined, revealing detailed patterns and enabling what has been termed dataveillance—the sorting and sifting of data sets in order to identify, monitor, track, regulate, predict, and prescribe.[33] The widespread generation of data and the practices of dataveillance raise many questions concerning privacy and rights to anonymity and confidentiality that are only just starting to be thought through and responded to.[34]

Data have long been used to profile, segment, and manage populations, but these processes have become much more sophisticated, fine-tuned, widespread, and routine with the application of data analytics employing machine learning techniques.[35] While the state might profile its citizens for the purposes of security and policing, commercial enterprises are seeking to reduce risk and maximize yield through more effective targeting of products. Whereas earlier generations of profiling sought to create aggregated population or area profiles, which then shaped decision making with regard to marketing and product placement (e.g., geodemographic profiling), new generation analytics can work at the level of the individual, combining data from various sources such as credit and store card transactions, clickstreams, social media posts, and other kinds of personal data to produce a detailed customer profile.[36] These profiles are used to socially sort customers, identifying some for preferential treatment and excluding others, and to predict the likelihood that customers might be able to meet payments or to judge their projected lifetime value if they remain loyal, and how likely they are to move their custom.[37] They are also being used to underpin new forms of dynamic and personalized pricing, tailored to a consumer's profile and purchase history, that are designed to leverage optimal spending.[38] Consumers are thus being routinely measured and ranked, and they receive differential services, based on their associated data and where they live.

One particularly pernicious form of predictive profiling is anticipatory governance. It involves predictive analytics that are used to assess likely future behaviors or events and to direct appropriate action. Such anticipatory governance has been a feature of air travel for a number of years, with passengers profiled for risk and levels of security checks prior to starting their journey.[39] More recently it has been extended to general policing, with a number of U.S. police forces using it to identify potential future criminals and to direct the patrolling of areas based on an analysis of historical crime data, records of arrests, and the known social networks of criminals.[40] In such cases individuals' data shadows do more than follow them; the data shadow precedes them, seeking to police behaviors that may never occur.[41] As a consequence, people are treated differently in anticipation of something they may or may not do. Given their effects vis-à-vis individual lives and their black-boxed nature, the practices of predictive profiling, social sorting, and anticipatory governance require much more attention, as do the companies that develop and undertake such tasks.

The work that data systems do in all of these cases is based on generating an excess of data. Indeed big data is premised on generating, hoarding, and linking as much data as possible in the hope that value and insight can be leveraged from them. Rather than being generated and used to fulfill a specific task, data can be repackaged, sold, and repurposed for all kinds of secondary uses. Such a strategy runs counter to the policy of data minimization, one of the foundations of privacy and data protection in the European Union and North America. This policy stipulates that data should only be generated and used to perform a particular task and that they should be retained only for as long as they are required to perform that task.[42] A clear example of where the premise of data minimization is being breached is with respect to control creep, in which data generated for one form of governance is appropriated for another.[43] Clearly control creep has mostly occurred with respect to security, with airline industry and government administrative data being repurposed for profiling and assessing passenger risk; with congestion charge cameras installed for that sole purpose also being used for general policing; and with social media data being repurposed to conduct criminal investigations and undertake predictive profiling.[44] But control creep is also in evidence across a range of other domains, for example, using personal location, consumption, and social media data to assess credit risk or suitability for employment.[45] Given the implications for civil liberties from secondary data use, there is a need to examine its consequences and to design new approaches to data protection, such as privacy by design.[46]

Conclusion

Dalton and Thatcher conclude their call for critical data studies by setting out five questions that they believe require further study, all relating to big data:

- What historical conditions lead to the realization of big data such as they are?
- Who controls big data, its production, and its analysis? What motives and imperatives drive their work?
- Who are the subjects of big data and what knowledges are they producing?
- How is big data actually applied in the production of spaces, places, and landscapes?
- What is to be done with big data and what other kinds of knowledges could it help produce?[47]

There are many more questions that can be added to this list, not least by widening the lens to open data, as well as data archives and repositories, but also by considering the wider data landscape, data assemblages, and data markets. Rather than produce an extensive list of questions, we want to conclude by calling for greater conceptual work and empirical research to underpin and flesh out critical data studies.

The ways in which data are being generated, the analytics used to process and extract insight from them, the industries growing up around them, their wider political economic framing, and how they are employed all demand critical engagement. While there is a rich and diverse tradition of critical social theory that can be directed toward data assemblages and the wider data landscape, such theory needs to be refined and fine-tuned to make sense of data and their work in the world, with new theory developed where needed. Yet we have barely begun to critically conceptualize data and their apparatus and elements. Such thinking needs to be complemented with more normatively oriented reflection on the ethics and politics of big data, open data, and data systems of different varieties.

Such conceptual and normative assessments need to be accompanied by a diverse set of empirical case studies that examine all facets of data-driven governance, business, and science, that unpack data assemblages, and that map the wider data landscape. Our suggested approach is to employ methods such as ethnographies, interviews, focus groups, and participant observation to delve into the workings of assemblages, to trace out genealogies of how the data landscape has changed over time and space, to map the materialities and infrastructures that constitute data infrastructures, and to deconstruct the discursive regime accompanying data-driven initiatives.[48]

Undertaking this conceptual and empirical work is what our own research will focus on over the next few years as part of the Programmable City project, building on our initial large-scale studies.[49] This extensive project is examining the intersections of big and open data, ubiquitous computing, software and algorithms, and smart city developments in Dublin and Boston, unpacking a set of data assemblages and charting the data landscape of each city. We have no doubt that many others will be engaging in similar studies, given the growth in data-driven forms of science, business, and government. We hope that what this research will produce is a diverse set of vibrant critical data studies.

Notes

1. Dupaquier and Dupaquier, *Histoire de la démographie*; Bard and Shubert, *Encyclopedia of the Archaeology*; Poovey, *History of the Modern Fact*; Porter, *Rise of Statistical Thinking*.
2. Poovey, *History of the Modern Fact*; Garvey, "*facts* and FACTS"; Rosenberg, "Data before the Fact."
3. Kitchin, "Big Data and Human Geography," 262; boyd and Crawford, "Critical Questions for Big Data"; Dodge and Kitchin, "Codes of Life"; Laney, 3D *Data Management*; Marz and Warre, *Big Data: Principles*; Mayer-Schönberger and Cukier, *Big Data: Revolution*; Zikopoulos et al., *Understanding Big Data*.
4. See, for example, Hilbert and López, "World's Technological Capacity"; Gantz and Reinsel, *Extracting Value from Chaos*; and Short et al., *How Much Information?*
5. Gantz and Reinsel, *Extracting Value from Chaos*; Manyika et al., *Big Data: Next Frontier*.
6. Rial, "Power of Big Data."
7. Constine, "How Big Is Facebook's Data?"; Open Data Center Alliance, *Big Data Consumer Guide*.
8. Lauriault et al., "Today's Data"; Kitchin, *Data Revolution*.
9. Floridi, "Data"; Rosenberg, "Data before the Fact."
10. Pérez-Montoro and Díaz Nafría, "Data."
11. Gitelman and Jackson, "Introduction."
12. Bowker and Star, *Sorting Things Out*; Lauriault, "Data, Infrastructures"; Ribes and Jackson, "Data Bite Man"; Kitchin, *Data Revolution*.
13. Gitelman and Jackson, "Introduction," 2, citing Bowker, *Memory Practices*.
14. Star and Ruhleder, "Steps toward an Ecology"; Kitchin and Dodge, *Code/Space*.
15. Ruppert, "Governmental Topologies"; Hecht, "Technology, Politics, and National Identity," 257.
16. Lauriault, "Data, Infrastructures"; Ruppert, "Governmental Topologies."
17. Kitchin, *Data Revolution*.
18. Dalton and Thatcher, "Critical Data Studies."
19. Kitchin, *Data Revolution*, 24.
20. Ribes and Jackson, "Data Bite Man."

21. Foucault, "Confession of the Flesh," 194.
22. Foucault, "Confession of the Flesh," 196.
23. Hacking, "Biopower"; Hacking, "Making Up People"; Hacking, "Tradition of Natural Kinds"; Hacking, "Philosophie et histoire des concepts scientifiques"; Hacking, "Kinds of People."
24. Hacking, "Tradition of Natural Kinds"; Hacking, "Philosophie et histoire des concepts scientifiques"; Hacking, "Kinds of People."
25. Lauriault, "Data, Infrastructures."
26. Hacking, "Philosophie et histoire des concepts scientifiques."
27. Hannah, *Dark Territory*; UK Census, *Irish in Britain*; Middleton, "Email Urges 'New Zealander'"; Waring, *If Women Counted*; Singler, "SEE MOM IT IS REAL."
28. Kitchin, *Data Revolution*.
29. Amoore, "Biometric Borders"; Bamford, *Shadow Factory*.
30. Kitchin and Dodge, *Code/Space*.
31. Koops, "Forgetting Footprints."
32. CIPPIC, *On the Data Trail*.
33. Clarke, "Information Technology"; Raley, "Dataveillance and Countervailance."
34. Solove, "Taxonomy of Privacy"; Elwood and Leszczynski, "Privacy, Reconsidered."
35. Weiss, *Clustering of America*; Goss, "'We Know Who You Are'"; Parker, Uprichard, and Burrow, "Class Places and Place Classes"; Singleton and Spielman, "Past, Present and Future."
36. Siegel, *Predictive Analytics*.
37. Graham, "Software-Sorted Geographies"; Minelli et al., *Big Data, Big Analytics*.
38. Tene and Polonetsky, "Big Data for All."
39. Dodge and Kitchin, "Flying through Code/Space"; Amoore, "Biometric Borders."
40. Siegel, *Predictive Analytics*; Stroud, "Minority Report."
41. Stalder, "Privacy Is Not the Antidote"; Harcourt, *Against Prediction*.
42. Tene and Polonetsky, "Big Data for All"; CIPPIC, *Submissions to the House of Commons*.
43. Innes, "Control Creep."
44. Lyon, *Surveillance Studies*; Pither, *Dark Days*; Gallagher, "Staking Out Twitter."
45. O'Reilly, "Creep Factor."
46. Information and Privacy Commissioner/Ontario, *Seven Principles of Privacy*.
47. Dalton and Thatcher, "Critical Data Studies."
48. Kitchin, *Data Revolution*.
49. See Programmable City, http://www.nuim.ie/progcity/; Lauriault, "Data, Infrastructures"; and Kitchin, *Data Revolution*.

2 Big Data ... Why (Oh Why?) *This* Computational Social Science?

David O'Sullivan

Many others, in this volume and elsewhere, have and will comment on the political, social, economic, cultural, and ethical implications of big data.[1] I strongly agree that those implications are important. Indeed I believe they are the most urgent aspects of big data with which critically engaged social science must grapple. There is a good fight to be fought in the public arena over the many worrying directions in which the political-economic impulses driving big data are pointing. The complicated ways in which the big data movement (if we can call it that) is entangled with a burgeoning, authoritarian, and surveillant state, simultaneously enabling and in thrall to a rhetoric of free-market "disruption," demand our close attention, if they are to be countered by more humane alternatives. There is of course a substantial literature that refuses to roll over before the technological triumphalism (or is it fatalism?) of work such as *Too Big to Know* and *Big Data*.[2] Many of the concerns raised by the current moment are ably dissected in *Digital Disconnect, The Filter Bubble, You Are Not a Gadget, Who Owns the Future?*, and *To Save Everything, Click Here,* among others.[3] Given the centrality of geographical data of one kind or another to the data deluge, it is surely important that geographers become more visible in this public conversation.[4] Substantial contributions, such as *Code/Space*, remain firmly academic in tone but nevertheless provide a foundation for future contributions that tackle more specifically spatial aspects of big data and its impacts.[5]

Recognizing the importance of these wider debates, I nevertheless want to focus on narrower methodological concerns. Whatever else big data has accomplished, it has placed quantitative and computational methods firmly on the social science agenda.[6] I welcome that development and see it as providing an opening for a more plural vision of geography and other social sciences. However, it is unlikely that opening will lead us anywhere new if we persist in understanding big data as primarily about the novelty of the data themselves. Data, however "big," are severely limited in how they represent *processes*. Given the centrality of process to developing any sophisticated understanding of how the world works, this is more than a limitation of big data. If understanding, explaining, and effectively intervening in the world are the goals, then we must ask questions about the style of computational social science we ought to be aiming for. Yet there is every sign that (over)excitement and hype around big data are in danger of causing us to lose sight of such matters. This would be unfortunate for both opponents *and* proponents of the potential of big data for social science; drawing attention to these issues is therefore my aim.

In the next section I set out my understanding of the epistemology of big data and suggest why big data has been so successful—successful, that is, as a widely adopted technology, not necessarily as a way to understand the world. From there I move on to consider a persistent dualism in how computational tools have been deployed in the sciences, namely a distinction between top-down, aggregate, or statistical approaches to explanation (among which big data can be placed) and bottom-up, emergentist approaches often associated with complexity science.[7] While these two traditions share substantial elements in their intellectual heritage, they yield sharply divergent perspectives on explanation, understanding, and prediction and suggest very different intellectual styles and methodological directions for computational social science. The two traditions also tackle the central issue of process very differently. The current openness to greater use of computers and (secondary) quantitative data is an opportunity for better, more effective social science that we are in danger of missing if the big data paradigm remains dominant. That danger has both scientific and ethical dimensions, returning the argument to the broader context considered at the outset.

The Mysterious Rise of Big Data

Big data has seemingly come out of nowhere, very quickly, but this is illusory. Iconic magazine covers on the topic, such as *Nature*'s "Big Data: Science in the Petabyte Era" and the *Economist*'s "Data Deluge," popularized the term "big data" but were testament to developments already well under way.[8] Even so, the speed with which such a media-detected (and inevitably amplified) "trend" has morphed into a prime directive for all of science has been surprising. Living in New Zealand until the end of 2013 somewhat shielded me from this juggernaut, but even there by early 2013 big data was unavoidable, as a series of discussions on the topic, sponsored by the Royal Society of New Zealand and National Library and broadcast by the state-funded Radio New Zealand, makes clear.[9] As has happened elsewhere, national science funding and infrastructure initiatives were quickly hitched to this latest, urgent strategic imperative, opening profitable opportunities for private companies building New Zealand's Ultra-Fast and Rural Broadband Initiatives. As has happened in many other jurisdictions, these developments are often explicitly connected to the parallel evolution of "smart cities."[10] The details of New Zealand's experience of the big data and smart cities movement are geographically, politically, and culturally specific, but the commonalities with what is unfolding elsewhere are striking.[11]

The New Zealand case immediately brings to the fore the oft-discussed question of what exactly it is that makes big data *big*. Certainly the volume of data generated in New Zealand would not qualify as big in many other contexts. The by now overly familiar three Vs—volume, velocity, and variety—supposedly definitive of big data were, appropriately enough, purloined from a business intelligence report.[12] They have proven insufficiently descriptive for many tastes, leading to an academic cottage industry proposing and debating additional attributes (preferably ones starting with the letter V). Rather than add to that debate, I consider big data to be primarily an epistemological stance.[13] That stance can be crudely sketched as a claim that, given sufficient data, the world can be known (if not completely, then well enough for any particular purpose) and that we are currently on the threshold of an era in which the technological capacity to assemble such complete data sets has (at last!) arrived.[14] As such, this moment heralds the realization of a dream of certain kinds of deterministic positivism.[15] This is of course philosophically extremely shaky ground, although it appears that epistemological difficulties are not a deterrent to adoption of the approach.[16] Where big data may be deployed, the point in many contexts is not, after all, to understand the world but simply to know it well enough to make a profit or, an even lower bar, to be plausibly able to claim that a profit might someday be made using insights gleaned from data.[17] Suffice to say, while I am, along with many geographers, intrigued by the possibilities such data sets may open up, I am unpersuaded by the grandiose claims made for big data. Understanding the world still demands that we carefully develop theories, consider the implications of those theories for what we expect to observe in the world, and subject those expectations to scrutiny through empirical observation, using multiple methods, only a few of which are enhanced by the dragnet of big data collection.

In spite of the unconvincing epistemological claims, how is it that this particular computational approach has come to dominate so much recent thinking about using computation to learn about the world? At least three answers come to mind. First, it (retroactively) justifies data collection that rests on questionable ethical and legal foundations. Never mind how we came to be in possession of these vast data repositories; just think about what we can do with them! A case of ends justifying means on a societal scale.

Second, the big data techno-social regime is feasible in a context where collecting more data became necessary as a matter of everyday business practice. It is not clear that what has emerged was deliberately planned by any of the leading protagonists. For example, Brin and Page, in an early paper on the Google search engine, note that "we expect that advertising funded search engines will be inherently biased towards the advertisers and away from the needs of the consumers," suggesting that building a data-gathering, advertising company was not their original intention.[18] However, in a business environment where building an audience took priority over the difficult task of selling services that were often not obviously useful, it was imperative for those services to be free at the point of use, leaving targeted advertising and the attendant surveillance as one of the few viable, sustainable business models.[19] This path is one premised on a financial speculation that the data will eventually pay off, but regardless of the eventual correctness of that speculation, once a company starts down this path more data about users can only be better, and available technology has made the assembly of vast data sets possible.[20]

Third, from the perspective of making a profit, there is little doubt that big data can work. Indeed from this perspective profit (or more generally, efficiency) is the only metric that matters: "The capitalist correlation imperative is clear: spurious correlation is fine, so long as it is *profitable* spurious correlation."[21] This leads to a stance on knowledge that is unconcerned with explanation: just as I don't need to understand how my phone works to use it, corporations and governments don't need to understand how or why the algorithms they use work to operate them or, at any rate, they don't until the algorithms plainly aren't working.[22]

Big Data as Method

So much for a schematic explanation of the rise of big data. In practical terms what does big data *as method* consist of? Given the loose way in which the term is used, it is not easy to pin this down, but given the salience of the quantity of data in the approach, the emphasis is on large-scale data analysis methods, of various statistical kinds. It bears emphasizing that this orientation immediately places data ahead of theory, since data and the world they are assumed to illuminate come before any consideration of the questions to be addressed. In any case speaking very generally, statistical models are fit to data to identify factors accounting for variations in the data. To be sure, a wider range of methods is available now than at any earlier time, including machine learning, data mining, and pattern recognition methods, alongside more exploratory approaches, particularly interactive visualization.[23] In an ideal case a single researcher, analyst, or small team of analysts might use exploratory visualization methods to develop familiarity with a data set. This in turn might prompt some ideas about the patterns to be found there, the methods most likely to emphasize those patterns, and from there the statistical models most suited to advancing understanding of the phenomena represented by the data set. For many scientists working today with much larger data sets than ever before, this is a reasonable description of how they would proceed. It is also not very new or different from how they would have proceeded in the past.

But data volume and velocity *do* matter. When data sets are very large and rapidly changing, then the scope for an exploratory approach is limited, since the computational demands are potentially limitless. Problems must be rendered tractable by *predefining* what constitutes a pattern of interest in terms of the data known to be available. In corporate or other environments where timely, "actionable intelligence" is prized, much more constrained, automated approaches are likely to prevail. In these contexts much of the decision making about what patterns to attend to must be delegated to the diagnostic statistics generated by whatever methods are deployed. In a visionary piece Stan Openshaw (with some excitement) anticipated the type of continuous monitoring that this approach implies, along with a major drawback: the identification of many spurious patterns and correlations.[24] Such considerations demand that the criteria specifying what is interesting and what is not be narrowed further, forestalling the open-ended search for patterns that might inspire the collection of extensive data in the first place.

It would be absurd to argue that there is no potential in more detailed, more frequently updated data for describing better how social change unfolds over time. Perhaps previously unknown social phenomena can be observed as a result of the improved temporal resolution in such data. Loose analogies with the advances made possible in other fields by the invention of the telescope or microscope do not seem completely misguided.[25] Particularly when big data is used in conjunction with other approaches, there are surely grounds for (guarded) optimism about its social scientific possibilities. In a specifically geographical context Miller and Goodchild identify some of what is exciting about this approach and point to interesting continuities with previous work in quantitative geography.[26] Geography's long-standing challenge of bridging from the local and particular to the global and general is central to their argument that the big data approach, when thoughtfully and carefully deployed, holds promise for a "data-driven geography," particularly in the abductive early discovery phase of research when the aim is to develop good ideas and candidate explanations.

Even so, a larger point here is that the methods associated with big data start from the aggregate level and deploy a statistical approach to identify relationships among data attributes in more or less traditional ways. The mode of explanation is inferential statistical, based on a constant conjunction model of causality, rather than on a realist, mechanistic, or process-oriented account.[27] Contemporary large data sets, particularly those that are frequently updated, give an impression of dynamism and by extension may be considered to offer us a rich representation of process. In truth this is little more than an impression. Large data sets, even frequently updated ones, embody no concept of process. Sometimes it is implied that the *velocity* of such data sets, their currency, and frequency of update somehow capture process. In fact most such data are simply rapidly updating snapshots of events. Nothing recorded in the data captures the processes or mechanisms that drive the changes occurring in the data. Process and change are thus rendered as "one damn thing after another," with no notion of process or mechanism in the data themselves.

Instead data impose rather rigid concepts of identity on people and places, reducing process from "becoming" to mere change in attribute values associated with otherwise unchanged social entities (whether individuals or institutions). This distinction, and the need for a different approach to data that taking process, time, and change seriously entails, has been a focus in the geographical information science community for many years.[28] It is hard to conceive of any means by which process can be retrofitted to big data *as data*. Rather it is in creative use and practice, through analysis from a theoretically informed perspective, that concepts of process are added to data. Ironically such explicitly theoretically informed analysis is one of the approaches most loudly eschewed by the more aggressive advocates of big data.

Bifurcated Computation

Regardless of its epistemological limitations and process blindness, big data is clearly ascendant, at least for now. For my present purposes it is instructive to consider big data as the latest development in the deployment of computation in commerce and government and to pay particular attention to *alternative* computational approaches to understanding the world that are not so strongly favored at present. The origins of modern computation lie in World War II.[29] That context saw computers and the closely related field of operations research applied to the solution of the practical production and logistical challenges of mounting modern warfare on a large scale.[30] Of particular importance to the development of computing were the demanding mathematical problems that arose in these contexts, such as code breaking, complex optimization problems, and the simulation of nuclear reactions.

From the origins of computers we can identify two broad applications. First are applications of computation to data sets too large for calculation by hand, to produce closed-form solutions to mathematically well-defined problems, using various types of numerical analysis. Such calculations rely on algorithms for manipulating large matrices, on interpolation and approximation methods, and on the mathematics of linear algebra and optimization.[31] While routine, such computation is more demanding of computing resources, particularly as data sets grow in size, when the associated computational requirements may grow with the square, cube, or even higher-order powers of the problem size.[32] In principle such calculations are not difficult, but they are computationally intensive. This domain of computation is associated with the more efficient management of logistical systems, the optimization of resource allocations in production systems, and the field generally known as operations research.[33] The big data phenomenon sits squarely in this tradition.

A second broad area of applications can be identified where computation is iteratively applied to prospectively simulate real or hypothetical systems. Here the applications themselves may not be very data intensive, but repeatedly iteratively performing (often) simple calculations over time and/or space leads to substantial computational demands in the aggregate. Such computation is deployed in simulating target systems of interest, in meteorology, product design, military applications (flight and combat simulators), and perhaps most familiar to a general audience, in computer gaming.[34]

It is useful to consider these two computational approaches as they have played out in a particular discipline (geography) to provide a more specific account of the differences between them. Geography's quantitative revolution witnessed its own somewhat related bifurcation, in the divergence between theoretical model-oriented methods and more pragmatic applications of inferential statistics to primary and secondary data, typified by the coverage of texts such as King's *Statistical Analysis in Geography*.[35] It was not long before the peculiar challenges of spatial data complicated and compromised the latter enterprise considerably.[36] Meanwhile, beyond a few specialized areas and subfields such as urban modeling, the model-oriented approach found only limited acceptance before the dramatic epistemological upheavals of the 1970s and 1980s. As Thrift notes, this "ghettoizing of complexity theory in geography was a tragedy, since the potentialities for much wider interaction were there."[37] The marginalization of complexity-oriented approaches in geography, within *quantitative* geography, is instructive, because it emphasizes the extent to which any attempt to map methodological approaches onto political or other predispositions is doomed to failure.

Even from a computational perspective the distinction I am drawing is somewhat artificial, since many of the same computational tools and algorithms are equally applicable to either big data or complexity science, the point of general-purpose computing being precisely its mutable, reprogrammable nature. The truth is more complicated and nuanced than any simple binary account would suggest. On the one hand, simulation depends on repetitive, often routine calculation not achievable by hand. On the other, applying closed-form solutions to *small* data sets can enable iterative and interactive exploration of many possible solutions and lead from there to the concept of a solution space and ultimately to a more nuanced understanding of the original problem. When numerical analysis is applied in this way it can transform the questions asked of data. This more exploratory stance toward data sets has initiated the trend toward interactive visualization of larger, more complicated data sets.

Two Cultures of Computation?

It is tempting to map these two computational styles (closed solutions and open-ended exploration) directly onto two cultural manifestations of computing: authoritarian, corporate, statist, big brother, big data on the one hand and liberatory, individual-empowering, personal computing on the other.[38] The dualism is deeply etched into many accounts of the history of computing.[39] The duality is particularly emphasized by self-styled "revolutionary" or "disruptive" Silicon Valley start-ups, deploying their own version of the cultural politics of the post-1960s counterculture.[40] These contradictions are brilliantly dissected by Turner in his book *From Counterculture to Cyberculture*, and many of the contradictory oddities of high tech's self-consciously liberal (often libertarian) yet conservative elite are entertainingly recounted in Borsook's enduringly relevant *Cyberselfish*.[41]

A similarly odd clash of cultures is evident in the contrast between a New Age holism in the language and iconography around chaos and complexity science and the more authoritarian, establishment, and business agendas both funding and consuming this science.[42] Thus one important center of complexity science has been the Santa Fe Institute (SFI), which according to Helmreich "is sometimes considered the good twin of Los Alamos, concerned with the technology of life, rather than the technology of death," and established in part through the efforts of George Cowan, a former director of the Los Alamos laboratory.[43] Or again: "most scientists at SFI are wary of any association with New Age movements," and yet books such as Kauffman's *At Home in the Universe* struggle (for the most part failing) to stay on the scientific side of a surprisingly fine line between New Age flakiness and detached scientific rigor when it comes to the more grandiose claims of complexity science.[44] Such odd intellectual (if not cultural) contradictions may be the inevitable outcome when reductive scientific methods, so successful at explaining phenomena that are timeless at human scales (e.g., the solar system, evolution), are applied to systems in which historical modes of explanation, with their attendant contingencies and chance events, have been predominant.

A simplistic mapping of the cultural origins of the two computational traditions under discussion onto particular political or economic agendas is plainly unsustainable. Science and technology studies in numerous fields have repeatedly and convincingly demonstrated the highly contingent nature of the relationships between technologies and the politics they embed and produce.[45] So while it is tempting to suggest that complexity-oriented, bottom-up modeling is inexorably associated with anti-authoritarian and more open approaches to knowledge, while big data, top-down, classificatory, and inferential statistical approaches are aligned with powerful interests, it is demonstrably untrue. There is nothing intrinsic to either approach that determines the ends to which they can or should be deployed. Closed-form calculation might be used to optimize the efficient production and equitable distribution of medical or other public services, while simulation can be (and almost certainly has been) used to explore possible strategies for the illegal invasion and occupation of another country.

Contrasting Computational Epistemologies

The lack of a one-to-one mapping from computational approach onto particular political or economic agendas notwithstanding, it should nevertheless be clear that these two broad approaches as distinctive scientific practices embed different thinking about process. They also each sit more easily with contrasting attitudes to the use of computation in furthering understanding of socioeconomic, political, and cultural systems. Indeed it is my argument here that we ought to *choose* which approaches to the use of computation are more likely to advance our understanding of the world and then adopt them for that reason. It is likely that such pragmatism would see big data deemphasized in favor of complexity models and other computational approaches more attuned to process and explanation, as in the digital humanities.[46] This contrast between complexity science and big data is schematically elaborated in table 2-1.

TABLE 2-1. Differing approaches to complexity science and big data

COMPLEXITY SCIENCE	BIG DATA
Theory embedded in models	Correlation and classification
Process	Temporal snapshots
Open-ended exploration of process implications	Exploration of already-collected data
Bottom-up orientation	Top-down orientation
Multiple levels and scales	Two levels (aggregate and individual)
Many alternative histories (or futures)	"Just the facts" (or optimal solutions)

Complexity-oriented, model-based approaches are precisely about process. It is a representation of process (however limited) that drives the dynamics of such models, and open-ended investigation of model behaviors can be considered an exploration of the conditions of possibility of the system being modeled. Before getting too excited about this, one should acknowledge that in many cases the notion of process embodied in such models is not significantly richer than that implied by the historical snapshots of big data. Change is most often cast in terms of changing attribute values of otherwise fixed and stable entities. Nevertheless the focus is on change, as well as the circumstances that produce change—a perspective that forces users of models to consider processes and mechanisms directly. Interesting model structures that combine both attribute change and systemic structural change are one possible advance in this regard.[47]

Open-ended exploration of dynamic models engenders a different, more humble, and more provisional attitude toward knowledge compared to predefining and then identifying "optimal" solutions or patterns of interest.[48] Simulative computation posits "possible worlds" (in the form of simulation results under different scenarios or model configurations) and implicitly acknowledges the speculative nature of the exercise.[49] Speculative computer modeling in this vein has led to the recognition, in some of the mathematical sciences (and more widely), of limits to knowledge and prediction, in the form of dynamic effects such as chaos, and of properties such as emergence, path dependence, positive feedback, and adaptivity, all of which are likely to preclude reliable prediction of a system's behavior over time.[50] The recognition of these system characteristics under the banner of complexity science calls, at least potentially, for greater humility on the part of scientists and a recognition of limits to knowledge *inherent in the nature of the systems under study.*[51] Understood this way, complexity science underwrites a pluralistic approach to knowledge that acknowledges the importance of understanding systems at multiple levels, from multiple perspectives, and using a variety of methods.[52] Recognizing that social systems are composed of complex individuals organized into households who play multiple roles in a range of institutions of varying organizational structures with a range of aims and goals, one ought to realize, when looking through a complexity science lens, that *no* single-level, top-down understanding of how society works will do. Furthermore, different methods are likely to be appropriate for getting at what is going on in each of these diverse contexts.[53]

It is important to note that big data and complexity science are not as far removed from one another as they at first appear. Both are about fitting simple models to observations: statistical models derived from observational data on the one hand and synthetic simulation models on the other. At the same time, the important differences in epistemology sketched here are real. A complexity-oriented modeling approach to knowledge allows us to think of data not as hard, precise evidence of reality but as a set of patterns that constrain a space of plausible, speculative models whose structure and mechanisms can account for those patterns and that may therefore be useful in building process-oriented, theoretical explanations for the existence of those patterns.[54] Data in this context become an intermediate step in the development of explanations. By contrast, under the model most often adopted in the world of big data, data themselves *are* the phenomena, and explanation is less about understanding processes and mechanisms—that is, *explaining* the world—and more about describing the data, at which point the phenomena themselves are taken to have been understood.

Conclusions

As I have tried to show, an orientation to process is absent from the epistemology of big data yet is surely central to any coherent approach to explanation in geography and the social sciences more generally. Other computational approaches offer more in this regard but have been less prominent in recent years. As much as anything this may be symptomatic of intellectual fashions in science. Chaos and complexity theory both had their own times in the sun, and "complexity" remains a much-trafficked buzzword. Perhaps these approaches were tried but failed to deliver on their initial promise, as I suggest big data is likely to do also. There may be some truth in this view, although it depends on an odd, fashion-conscious perspective on how we should evaluate scientific method. More seriously it fails to appreciate how great a challenge conventional modes of scientific explanation face in taking seriously the uncertainties introduced by complexity. The complexity enterprise points to a much wider remit for historical and narrative modes of explanation, which go against the grain of dominant modes of scientific explanation. The tension is exacerbated by the "simple rules, complicated behavior" mantra so often used to sell complexity science, which presents simple models as the end point of complexity-oriented approaches when really they are only the beginning. Just as it is foolish to believe that mining big data can provide answers to every social science question, it would be foolish to argue that simple complexity science models can answer every question.

Recognizing and valuing pluralism in methods is key to the complexity-oriented computational approaches I favor. That implies two things. First that there is of course a place for big data.[55] It would be absurd to argue otherwise. Without doubt when contemporary data sets and data-mining methods are applied to questions of genuine social scientific interest, new phenomena will be identified and new perspectives on old questions will emerge. But *understanding* those new phenomena will demand approaches other than those of big data. This leads immediately to the second point—

that other approaches to geography and social science remain vital to any coherent way forward. What is disturbing about much of the hype around big data is the apparent desire to advance on all fronts simultaneously: big data, not content with being a revolutionary approach to social science, must become a whole system for living, a societal lifestyle choice, a new mode of governance, of business, and of science.

The tragedy is that this stance toward big data could easily discredit all computational approaches to the social world—not only by getting the science wrong but also by becoming a pervasive social surveillance system, the necessity of which is unclear, beyond the desire for profit of large corporate interests and the data anxiety of a surveillant state.[56] The psychology of big data holds out an entirely false promise: that if only the data were bigger, we would know even more. There are certainly contexts where this might be true (astronomy's Square Kilometer Array, for example), but they are not social ones.[57] Much of what is revealed by social big data we either already know or can access in other ways that can place human actions and decisions in much richer social contexts. It seems likely that we would lose very little of genuine scientific interest by *not* recording and storing every person-machine microinteraction. There are no scientific grounds for "collecting it all," only commercial imperatives (and even those are founded on a wild speculation), which returns us to the important political-economic issues mentioned at the beginning of this chapter.[58]

There is a danger in focusing as I have on method, on emphasizing means over ends. It is ethical positions, not methodological choices, that most affect the impact of research.[59] Ultimately the two cannot be disentangled, and if the undoubted potential of computational methods in the social sciences is to be realized, it is important that we discover what can be learned from past mistakes, recognize the limitations of all our data, and focus on developing computational approaches to geography and social science better aligned with handling those problems. Such a social science will not only be better science *qua* science but will also be more ethically defensible as a direct result of recognizing the explanatory limitations of data. In sum we should not only be challenging the political economy of big data but ought to be deeply (and vocally) suspicious of its epistemology, not only from within critical traditions skeptical of quantification anyway but from the perspective that more interesting quantitative and computational methods are available.

Notes

1. "Big data" is a troublesome phrase to use correctly given its willful (and annoying) mixing of a singular adjective and a plural countable noun—"numerous data" would have been a more accurate, if less compelling coinage. Further, the phrase "big data" has come to stand for a complex mix of technologies, ideas, and practices, such that it may be considered a singular noun phrase. In a perhaps futile attempt to hold back the tide of offenses against grammar, where I consider the meaning to be "many data" I treat "data" as plural, whereas when the meaning is big data as an idea, I treat it as singular.
2. Weinberger, *Too Big to Know*; Mayer-Schönberger and Cukier, *Big Data: A Revolution*.
3. McChesney, *Digital Disconnect*; Pariser, *Filter Bubble*; Lanier, *You Are Not a Gadget*; Lanier, *Who Owns the Future?*; Morozov, *To Save Everything*.
4. Farmer and Pozdnoukhov, "Building Streaming GIS cience."
5. Kitchin and Dodge, *Code/Space*.
6. Barnes, "Not Only . . . But Also"; Barnes and Sheppard, "Nothing Includes Everything"; Burrows and Savage, "After the Crisis"; Johnston et al., "Mutual Misunderstanding and Avoidance"; Wyly, "Strategic Positivism."
7. Coveney and Highfield, *Frontiers of Complexity*; Manson, "Simplifying Complexity"; Mitchell, *Complexity: A Guided Tour*; O'Sullivan, "Complexity Science and Human Geography"; Thrift, "Place of Complexity"; Waldrop, *Complexity: The Emerging Science*.
8. These articles appeared in *Nature*'s September 2008 issue and in the *Economist*'s February–March 2010 issue.
9. Radio NZ, http://www.radionz.co.nz/national/programmes/bigdata, is where the series was archived at the time of writing.
10. For information about the infrastructural investments, see "Fast Broadband: New Zealand's Internet Upgrade," Ministry of Business, Innovation, and Employment, http://www.med.govt.nz/sectors-industries/technology-communication/fast-broadband. Some sense of the high priority given to big data in national science and innovation priorities is provided by the reports of the New Zealand Data Futures Forum, http://www.nzdatafutures.org.nz/.

11. As Kitchin, *Data Revolution*, reminds us we should expect.

12. Laney, *3D Data Management*.

13. boyd and Crawford, "Critical Questions for Big Data."

14. Anderson, "End of Theory."

15. Wyly, "New Quantitative Revolution."

16. Pigliucci, "End of Theory in Science?"

17. Wyly, "Automated (Post)Positivism"; Wilson, "Location-Based Services."

18. Brin and Page, "Anatomy of a Large-Scale Hypertextual Web," 18.

19. Some insight into how such intentions drifted is provided in Zuckerman, "Internet's Original Sin."

20. Wilson, "Location-Based Services."

21. Wyly, "Automated (Post)Positivism," 681 (original emphasis).

22. See Flood, *Fires*; and the earlier Wallace and Wallace, *Plague on Your Houses*, on which Flood draws.

23. See Miller and Han, *Geographic Data Mining*, for a survey of these developments in geography.

24. Openshaw, "Two Explanatory Space-Time-Attribute Pattern Analysers."

25. Brynjolfsson and McAfee, "Big Data Boom."

26. H. Miller and Goodchild, "Data-Driven Geography."

27. Sayer, *Method in Social Science*.

28. O'Sullivan, "Geographical Information Science."

29. Ceruzzi, *History of Modern Computing*; Dyson, *Turing's Cathedral*.

30. DeLanda, *War in the Age of Intelligent Machines*, offers a critical historical overview, while Shrader, *History of Operations Research*, gives a feel for the scope in a three-volume official history.

31. Aspray and Gunderloy, "Early Computing and Numerical Analysis."

32. Cormen, *Algorithms Unlocked*.

33. Light, *From Warfare to Welfare*.

34. Crogan, *Gameplay Mode*.

35. Chorley and Haggett, *Models in Geography*; King, *Statistical Analysis in Geography*.

36. Gould, "*Statistix Inferens.*"

37. Thrift, "Place of Complexity," 60n2.

38. It is particularly tempting if you live in Northern California and perhaps even more so if you have only recently moved there!

39. See, for example, Barbrook, *Imaginary Futures*; and Levy, *Hackers*, or consider the Apple ad broadcast during the 1984 Super Bowl; see Golumbia, *Cultural Logic of Computation*.

40. Frank, *Conquest of Cool*.

41. Turner, *From Counterculture to Cyberculture*; Borsook, *Cyberselfish*.

42. Thrift, "Place of Complexity."

43. Helmreich, *Silicon Second Nature*, 43. Los Alamos is of course home to one of the United States' national laboratories most closely associated with past and ongoing development of nuclear weapons; see Waldrop, *Complexity*, 53–69.

44. Helmreich, *Silicon Second Nature*, 41; Kauffman, *At Home in the Universe*.

45. Feenberg, *Critical Theory of Technology*; Latour and Woolgar, *Laboratory Life*; Winner, *Whale and the Reactor*.

46. Burdick et al., *Digital_Humanities*.

47. Gross and Blasius, "Adaptive Coevolutionary Networks."

48. Of course model-building science is not always an open-ended exercise. A considerable amount of economic theory is built on reversing the sequence of model building and exploration leading to the refinement of testable theory. Instead models provide rhetorical support for theoretical positions already firmly established, and they are refined to fit theory, rather than the other way around. In the process the models become more, not less, divorced from reality; see Lawson, *Economics and Reality*; and Keen, *Debunking Economics*. I am concerned here with an approach to the use of simulation models, discussed in my recent book; see O'Sullivan and Perry, *Spatial Simulation*. The proper use of simulation in science is a difficult philosophical area, which remains underexplored by philosophers of science, but see Winsberg, *Science in the Age of Computer Simulation*; and Weisberg, *Simulation and Similarity*.

49. Casti, *Would-Be Worlds*.

50. Gleick, *Chaos*.

51. Coveney and Highfield, *Frontiers of Complexity*; Manson, "Simplifying Complexity"; O'Sullivan, "Complexity Science and Human Geography"; Cilliers, *Complexity and Postmodernism*; Richardson, Cilliers, and Lissack, "Complexity Science."

52. Harvey and Reed, "Social Science."

53. Manson and O'Sullivan, "Complexity Theory in the Study of Space and Place."

54. Grimm et al., "Pattern-Oriented Modeling."

55. Miller and Goodchild, "Data-Driven Geography."

56. Lazer et al., "Parable of Google Flu"; Crampton, "Collect It All"; Crawford, "Anxieties of Big Data."

57. See Taylor, "Square Kilometre Array."

58. Crampton, "Collect It All."

59. Lake, "Methods and Moral Inquiry."

PART 2
Methods and Praxis in Big Data Research

3   Smaller and Slower Data in an Era of Big Data

Renee Sieber and Matthew Tenney

Big data has become a paradigm for a supposed new type of data and a form of knowledge discovery.[1] The hyperbole, if not necessarily the actuality, that surrounds big data now has an impact on nearly all aspects of our social life and increasingly has altered the way researchers and practitioners work with digital information. The language of big data dichotomizes our work as new versus old and big versus small. It affects the primary subjects of our inquiry, furthering the roles of people as sensors, communication as content, and living as behavior. The impacts are numerous, for example, with lucrative markets for data brokerage that supply the demand for a "data-now" business environment.[2] Governments, from the municipal to national levels, are transitioning from old to new ways of administering public services and resources via, for example, data analytics.[3] Skepticism accompanies public and private sector initiatives as big data passes "the top of the Hype Cycle, and [is] moving toward the Trough of Disillusionment."[4] Critiques within academia attempt to clarify the bewildering claims of big data by exposing its varied theoretical pitfalls and practical shortcomings to produce meaningful or actionable results.[5] Recent literature does little to dent the exaggerated hopes for big data, even as these hopes persist within the academy. As Wilson notes, social scientists are often quick to mistake big data as being a window into the daily life of its creators, claiming that "for many [of the general public], tweeting, posting, retweeting, and sharing is akin to breathing."[6]

We were drawn to big data for a pragmatic reason. An author of this chapter submitted a journal article focusing on Web 2.0 mapping platforms applied to local community development. According to a well-meaning reviewer, the number of contributions (n < 100) was deemed too small and too slowly contributed to offer the rigor expected in making informed statements about community needs or desires. We argue that this represents an increasing tendency to frame geospatial technologies and diverse approaches for conducting social science research such that they conform to approaches related to big data. Social processes are thus articulated in terms of volume and take place on cloud-based platforms. This type of data determinism, in which technology dictates the manner in which we understand society, can lead to a new form of publication bias. Properties of the specific technology overrule the value of very small and slow data and revive periodic arguments within geography on quantitative over qualitative methods.[7]

This chapter is not a postpositivist critique upon which quantitative methods of big data rely. Indeed we are agnostic about the value of big data and data-science methodologies. And we are not completely disheartened by the increasing role digital data has in social science inquiry, a phenomenon that can be studied and, *contra*, also can evince a phenomenon.[8] Our concern is that, by casting all data research in relation to big data, data scientists exacerbate internal issues within disciplines like geography and assist the momentum of a data-hungry social science. Framing data as needing to be big (e.g., large volume and high velocity) engenders expectations of validity and truth and creates a normativity in social science of how research "ought to be" conducted. We address these issues by first discussing the multiple origins of the concept of big data, the emergence of small data in response to the failings of big data, and the plight of very small and slow data as it is compelled to complement the epistemology of big data.

Big Data, from Industry Purpose to Industrious Purpose

According to Hidalgo, little "hype has come from the actual people working with large data sets. Instead, it has come from people who see 'big data' as a buzzword and a marketing opportunity—consultants, event organizers and opportunistic academics looking for their 15 minutes of fame."[9] It is important at the outset to state that there is no unifying definition of big data. Instead "big data" as a term has diverse origins and serves multiple needs stemming from academia, industry, and media. Mashey was perhaps first to explain the concept before the term became common.[10] He talked of "infrastress": vast amounts of data were placing excessive demands on existing computing infrastructures and required new approaches for data storage, database structures, instant access, and analytic services. To Mashey, big data offered no newly opened window into the social or customer world.[11] Rather big data presented a technical challenge to hardware and software engineers. Weiss and Indurkhya used the term in the name of a chapter in their book *Predictive Data Mining: A Practical Guide*.[12] The authors, with backgrounds in artificial intelligence, viewed big data handling as an opportunity that was not without inherent challenges: "Very large collections of data . . . are now being compiled into centralized data warehouses, allowing analysts to make use of powerful methods to examine data more comprehensively. In theory, 'big data' can lead to much stronger conclusions for data-mining applications, but in practice many difficulties arise."[13]

What began as a data management issue soon turned into a novel way of thinking about data. Tony Hey and fellow researchers at Microsoft went so far as to proclaim the emergence of a "fourth paradigm" of science, in which big data called for new methods to address the "current scientific data deluge."[14] Elwood, Goodchild, and Sui explicated the three prior dominant paradigms of twenty-first-century science: "the empirical (by describing natural phenomena), the theoretical (by using and testing models and general laws), and the computational (by simulating complex phenomena using fictional/artificial or small real-world data sets) approaches."[15] That data deluge, according to Chris Anderson of *Wired* magazine, heralded the end of theory.[16] Current theory-driven models were inadequate and inappropriate in the face of the opportunities afforded by volume. We no longer needed to rely on hypotheses to deduce relationships in complex systems.[17] Instead society required a data-driven science to explore and discover relationships where none were previously known to exist. According to Jim Gray, another Microsoft employee, data-driven science was akin to a "macroscope"—a situation in which research problems were investigated as though one were peering at millions of interactions through a microscope.[18] An inductive, data-intensive approach to science would "serve as the new ark upon which we can survive the current big-data deluge."[19]

The argument was that big data fundamentally differed from a very large data set and that this difference demanded new approaches that broke from traditional norms of science. In the data deluge precision has been offered as a replacement for accuracy, which cannot be evaluated because there are too many data. Big data sources such as Wikipedia utilize crowdsourcing to refine data.[20] The most common articulation of the difference between very large data sets and big data came from Laney, who reflected on his first-hand experiences with the challenges of big data storage and management. Laney characterized this new type of massive data set through what became known as the essential three *V*s. He argued that data sets were far more voluminous than before; there were new platforms on which data continuously streamed or were available at much higher velocities.[21] Data also were now accessible and analyzable as individual records, as opposed to entire data sets. Many of these data now manifest in highly unstructured and messy forms. Volume, variety, and velocity posed challenges in data integration and system interoperability. Today diverse accounts of big data are occasionally accompanied by a fourth, fifth, and even sixth *V* as necessary supports for the utility of specific big data applications (e.g., veracity, viability, variability, value). It is with the challenges (and opportunities) of these *V*s that big data features began to substitute for a singular definition, which we illustrate with an equation:

big data = f(volume, variety, velocity, and maybe veracity, viability, variability, value . . .)

Big data has solidified around these first three features of volume, variety, and velocity. Difficulties in characterizing the phenomenon of big data nonetheless remain. A recent informal survey by the School of Information at the University of California, Berkeley, asked "more than 40 thought leaders in publishing, fashion, food, automobiles, medicine, marketing, and every industry in between how exactly they would define the phrase big data."[22] Unsurprisingly their results revealed nearly forty unique definitions of big data that largely revolved around a particular contributor's application needs. What becomes evident after just a few entries is that big data has become a colloquial phrase that valorizes the potential of realizing granular insights relevant to specific goals rather than the mythical access to a generalizable data set that can be leveraged for numerous unanticipated uses.

According to Floridi, many of the definitions of big data rest on ambiguity and circular reasoning: data is big only in relation to our current computational power.[23] As M. Graham and Shelton put it, the "modifier 'big' [is] always relative and represents a moving target."[24] Contemporary "small data" were extraordinarily large a half century ago, and contemporary notions of big data will likely be tiny just a half century into the future. Batty argues that conceptual ambiguity has led to a joint focus on creation and use of big data instead of concentrating merely on volume.[25] Just being big (e.g., larger than an Excel table) does not render data valuable. Big data therefore remains "an abstract concept" that is only set "apart from masses of data, [by] other features, which determine the difference between itself and 'massive data' or 'very big data.'"[26] Additionally big data becomes constitutively inextricable from the capabilities of available software and hardware of the day.[27] Thus if it exceeds the capacity of a spreadsheet, then perhaps it is still big. The irony is that big data is not entirely the product of machines. Big data can trace its origins to the period from the 1880s to the 1940s at the Harvard College Observatory, where one-half million observations of the night sky which were amassed entirely by humans.[28]

Despite ambiguities or perhaps because of them, big data quickly transitioned from a term used to describe data collection and management to a marketing slogan that promised to enhance business practices and target customers.[29] The slogans shifted from "Big Data as Boogeyman," signaling that early misgivings over costs trumped the potential value of big data, to "The Big Data Gold Rush," in which big data was credited with creating a data market worth $125 billion.[30] The emergence of big data as a commercial industry clarified big data as a commodity that could be leveraged for business intelligence and as a subject for data science that could lend firms a competitive advantage. Big data also foregrounds the role of information technology (IT) in businesses, from a function the firm relegated to an IT department (e.g., payroll, inventory, and projections) to a core function in which an agile firm responds quickly to constantly changing data.[31] Agility requires new IT investments to manage data flows through, for example, new analytics, visualizations, and user interfaces. In this way data as marketing slogan returns to the original realm of big data in computer science.

Early marketing campaigns were paired with phrases such as "data overload" or "infobloat," which portrayed big data science as the solution to excessive and unwieldy content.[32] Outside a few anecdotal cases, big data continues to fail in delivering on the insights and value. A survey of more than three hundred IT departments found many big data–oriented projects never left the planning stages, with the proofs-of-concept and prototypes failing to reap value for their firms.[33] The same survey found a lack of empirical knowledge has already resulted in costly mistakes. Survey results mirrored an increasing disillusionment in the business community. News sites like ZDNet and Forbes suggested that big data was "oversold" relative to results: "Big data is hard (and the domain of the few). Doing it at scale and waiting for trickle down benefits can take time."[34] This has contributed to the ambiguity in big data definitions. Growing amounts of digital data are viewed as a panacea for industries and scientific endeavors while being accompanied by deflated expectations. To achieve the value promised by big data, it appeared we needed to bring in more data under its umbrella.

From Big Data to Small Data

Despite the promises of analytics or data visualization Eureka moments, big data was failing to extract value. It was too big, too fast, and too heterogeneous; it was incomprehensible and impersonal. "While companies (and computers!) like big data, most people only need small data," asserted Fidelman, because, compared to big data, "it is easier to analyze and test small data sets to differentiate signal from noise to extract meaning."[35] Segments of big data were increasingly seen as a cure for big data's lack

46   *Sieber and Tenney*

of utility: "one good strategy to solve the 'curse of big data' . . . is the intentional and purposeful breakdown of large data sets into smaller data sets."[36] A definition soon emerged from the private sector to formalize the concept of small data, which "connects people with timely, meaningful insights (derived from big data and/or 'local' sources), organized and packaged—often visually—to be accessible, understandable, and action- able for everyday tasks."[37] Small data echoed similar ambiguity found in big data, but small data was viewed as a way to deliver on the promises of big data without inducing extraordinary effort to extract value.

We argue that the concept of small data emerged for two reasons. First, small data offers a way to derive value from data sets using the same data science and analytics designed to reveal value in big data. Second, small data asserts the primacy of big data in framing all data. We would have no small data without big data, because "prior to 2008, data were rarely considered in terms of being 'small' or 'big.' All data were, in effect, what is now sometimes referred to as 'small data' regardless of their volume."[38] We turn to recent literature for several perspectives of the emergence of small data vis-à-vis big data. The emergence of small data helps us construct an epistemology of big data but at some cost to the integrity of small data:

small data = big data-some data

The first perspective is that small data is merely a digestible chunk of big data. Timely and meaningful insights derive from the deliberative extraction of subsets. Often this smaller data set is extracted because it responds to a particular organization or need.[39]

The initial process conducted on any big data is to reduce the data set in some meaningful way.[40] The process represents both a reduction and a recognition that "data in the wild" is never raw.[41] The reduction of big data minimizes the cost of data handling and presumably then maximizes the insights from otherwise bloated data sets. That is, utility is achieved by sampling and removing redundant, erroneous, and irrelevant data. In that way we produce working data sets. According to Lu and Li, data scientists rarely conduct analyses on big data; they ultimately utilize small data: "Most of the time, the direct access to the entire data is neither possible nor computationally feasible, forcing people to probe the properties of the data by looking at a sample. Because of the huge size of the data, quite often even a sufficient sample is too costly to obtain considering the network traffic involved and daily quota imposed. For practical consideration, we are often limited to the smallest possible sample."[42]

Jacobs provides an example of the computational challenges of handling big data.[43] The researcher generated a synthetic database consisting of 6.75 billion 16-byte records that was intended to emulate a censuslike record (e.g., age, religion, income, and address) for each person on the planet. The value of such a data source would be undeniably useful to geography researchers and others, and it was easy to store the records for the world's entire population on a single consumer-grade laptop in 2009. Jacobs argued that data storage does not present the limiting factor; analysis space is the challenge.[44] To derive insights (information) from massive number-crunching analyses, particularly when those data have temporal and spatial dimensions, requires the data scientist to respect the "aggregat[ion of] data in an order-dependent manner (for example, cumulative and moving-window functions, lead and lag operators, among others)."[45] The random access of most big data analytics destroys the temporal and spatial contexts of the data. Small data can maintain topology where big data could not.

Jacobs's example illustrates a recurring contradiction to amassing large data sets. We acquire the data even as we fail to amass the concomitant technological resources to handle such troublesome "bigness," and we may not acknowledge the uneven access to such big resources. Floridi clarifies the paradoxical challenge in which value from big data merits "more and better techniques and technologies, which will 'shrink' big data back to a manageable size."[46] Thus this first formalization of small data emphasizes the distillation of data from larger counterparts to avoid existing computational limitations and analytic overload. Bigness realizes value only when it becomes small. Even as our storage capacities grow, we will still likely need to "chunk" the data so we can analyze it. Here small data becomes a datum of big data, which is groomed by machines to the needs of an individual actionable effort:

Sieber and Tenney

```
small data <= human brain
```

A reason for failure of many big data projects can be attributed to a decision paralysis in the presence of all the possible tools, data sources, and potential applications available to big data.[47] The tools are essential, since incomprehensibility is considered an intrinsic characteristic of big data. A second perspective on small data refers to its capacity to improve on understanding big data.[48] Here small data is cast as data that is small enough in size for human comprehension.[49]

A working paper by Markowsky uses this humancentric definition of small data to justify human intervention in the subsetting of big data so that it "can be easily grasped by the human mind and easily visualized by the human eye."[50] Small data in this perspective is similar to the above description in the attempt to render data into familiar and manageable small data models. Instead of relying purely on a technological solution to derive big data insights (i.e., through computational analytics), this perspective embraces a traditional approach to interpreting data. The human brain becomes the analytical computer rather than depending solely on algorithms or statistical correlations crunching otherwise incomprehensible data sets.[51] The purpose of this characterization of small data is to aid people in using big data, so they can derive information and establish their own insights. This process proves difficult to not only replicate and intelligently subset.[52] It is also difficult to share, as we discuss below in the section "When Small Data Isn't the Answer, Regardless of Size.

```
small data = big data ∩ me
```

A different narrowing of the digital deluge has small data identified as that which is only about yourself. We find this perspective in the realm of the quantified-self movement, which comes from the rise of wearable and mobile technologies.[53] Devices such as Fitbit create relatively large volumes and velocities of content about individuals. These are the digital traces generated by nearly all aspects of technology we use, which can be in turn analyzed to derive insights about our own individual behaviors. Estrin and the Small Data Lab at Cornell University consider this type of data to be small data, which "we can think of . . . as [a] new kind of medical evidence, evidence where n = me, because it complements traditional big-N population studies with data that are just about me (or you) over time."[54]

Compared to big data, *big data ∩ me* is intended to be neither anonymous nor aggregated. It is intended to be comprehensible because it concerns a specific individual and mirrors the attempts to value big data by discretizing the data into digestible chunks. Small data thus acquires a personalized characteristic absent the collectivity of big data. This small data represents a new, highly personal source of valorizable information, where data reaches deep into the body, for example with embedded WiFi-connected medical devices like pacemakers. *Big data ∩ me* reveals an underlying moral quandary for small data. Individuals may generate the data, but this type of small data is largely out of reach for most individuals to obtain or effectively use and is further obfuscated by (lack of) rights to data ownership and privacy considerations.[55]

In the same way small data represented a break from big data in terms of distillation and comprehension, this perspective highlights a distinction in who the user is. In big data the end user equals analyst. In small data the end user can be the source of data (or collector, as we will see with national censuses) but not necessarily the analyst. It is likely the small data of the quantified self becomes aggregated across individuals for an analyst. The analytics and visualizations are built from the aggregations, which are then customized to an individual's data stream. Neither the devices nor the software would be developed were it for a single individual. Thus we begin to see the inextricable interplay between big and small:

small data = big data/domain

A fourth perspective on small data refers to big data shrunk by specific domains like geography. It is often related to data about the self: "the data on my household energy use, the times of local buses, government spending—these are all small data."[56] Importantly it also depicts a domain, for example, of energy, transportation, and public administration. Many of these feeds contain explicit (e.g., bus locations) or implicit (e.g., gov-

Sieber and Tenney

ernment spending, which is jurisdictionally bound) geolocations. At the 2012 meeting of the American Association of Geographers there was a special session entitled "Whither Small Data? The Limits of Big Data and the Value of 'Small Data' Studies," which led to a special edition of the *GeoJournal*.[57] At this session Goodchild and Kitchin characterized geographic data under the category of small data. Their primary example is the national census because its volume resembles a common characteristic of big data and because of the central role a census plays in many geographic inquiries.[58] Goodchild reinforces the perspective that small data is domain based as opposed to volume based. He argues that, in just the space of two years, small data has evolved from acting as proxy for big data to being a general term that situates the "traditional geographic approach" within the practice of small data studies: "Big data is distinguished from what I propose to term small data by its lack of the normal processes of quality control, documentation, and rigorous sampling. . . . Small data, exemplified by the products of the census, has supplied all of those things, with the result that analysis of small data readily leads to generalization."[59]

Here big data is either reduced to a specific domain or big data becomes more comprehensible when it is rendered amenable to specific domain methods. In a later piece H. Miller and Goodchild assert the value that geography brings to big data.[60] They argue that geographers possess lengthy experience with data volume (e.g., with Landsat remotely sensed imagery), as well as data velocity and variety (e.g., with volunteered geographic information [VGI] of multimedia geolocated content from social media platforms). Traditional methods were developed throughout the quantitative revolution in the field, survived the cultural geography backlash, and flourished in the GISCIENCE backlash to the backlash. These rendered the discipline as being arguably better prepared than some others to engage with big data and fuse that engagement with smaller social science research.[61] Traditional methods could be applied to newer geography data, such as VGI, which resembles big data in its "messiness," is "unstructured, collected with no quality control, and frequently accompanied by no documentation or metadata."[62] This perspective on small data opens up traditional geographic data sources, such as a national census, to new analytic techniques and new data, like VGI, to traditional geographic methods.

The implication is that disciplines can assert their relevance by transforming big data—"taming" data—into more meaningful data. In turn a specific knowledge domain gains relevance by its association with a new source of valorization. By positioning a discipline's data in relation to big data, a discipline is shown to be equipped to tackle a new data source and to be sufficiently important to be heeded by other disciplines. The value of this positioning vis-à-vis big data within geography is energized by claims for its powerful and unique ability to peer into layered and complex social systems. These claims are advertised by statements like this: "imagine, for example, the human geography and broader social science research that could be undertaken with the data set put together by President Obama's team for his 2008 and 2012 election campaigns."[63] Hyman points out that media speculation artificially elevated the electoral data–crunching techniques utilized, which was accomplished with relatively small data capable of being analyzed with paper and pencil.[64] This mirrors the hype in the private sector about the promise of knowledge discovery that can combine domain expertise with big data and data-driven science. We desire to see the potential for big data and its analytics even when it may not exist.

Goodchild has posited small data as data with quality control, documentation, and rigor.[65] This mirrors Kitchin and Lauriault, who offer a formalized definition for small data in which "small data are ... characterized by their generally limited volume, non-continuous collection, narrow variety, and are usually generated to answer specific questions."[66] There is a general lack of distinction in this usage of small data—between data and information—or what role data have in the various techniques of collection, analysis, and use within current geographic research. Despite this omission, the authors illuminate a critical difference between big and small data: most data prior to the vaunted bigness were targeted and organized with intent. Small data is goal-driven data, created with its specific goals and objectives. We will argue later that this intent, one of several distinguishing features of small data, can evaporate and thus

Sieber and Tenney

damage the defense of small data. More significant for us, this definition cements the intrinsic ties between small data and big data, as the former is defined with the modifiers of the latter and, as acknowledged by the authors, is susceptible to big data's science and practices:

> First, despite the rapid growth of big data and associated analytics, small data will continue to flourish because they have a proven track record of answering specific questions. Second, the data from these studies will more and more be pooled, linked, and scaled through new data infrastructures, with an associated drive to try to harmonize small data with respect to data standards, formats, metadata, and documentation, in order to increase their value through combination and sharing. Third, scaling small data exposes them to the new epistemologies of data science and to incorporation within new multi-billion data markets being developed by data brokers, thus potentially enrolling them in pernicious practices such as dataveillance, social sorting, control creep, and anticipatory governance, for which they were never intended.[67]

The prior definition highlights one last hoped-for perspective on small data—that small data is not related to big data but still serves as an input parameter of big data analytics:

small data ≠ big data, but value = big data analytics(small data)

Our central conclusion is that these varied small data perspectives, rather than offering different lenses on data, instead reassert the discourse on big data. Small data is more comprehensible, possesses more rigor, and so on, especially when those data are about us. However, that small data is also positioned vis-à-vis big data to presumably reap all of big data's advantages. As soon as we attempt to define big and small data we expose a circular problem: big data finds value only when made small, but small data, according to some, achieves value only when it is reassembled into something resembling big data. We basically roll around and around between big and small data and consequently gain no greater clarity on either type. Small data perspectives also call attention to long relationships between a domain like geography and Internet-related technologies. Small data as georeferenced data represents a separation from and an insertion into a big data epistemology in which a census may embed purpose to the data set but the multiple *V*s begin to matter for all kinds of data. In these inclusions of nominally big data into small data the proponents of the definition also recast geography by separating previous geographic data and practices from the tenets of future data. All data, particularly the eminently mashable geographic data, become part of big data.

When Small Data Isn't the Answer, Regardless of Size

Increased power and control, disruption, and new insights often accompany concepts of big and small data. Haklay, Singleton, and Parker identify how neologisms, especially those associated with the Internet, are common in many research fields that attempt to invoke legitimacy alongside dominant research agendas.[68] For us, neologisms serve as a shorthand for epistemology, a way of achieving truth that, for big data, lies in its capacity to be valued (e.g., monetized). By definition most neologisms are benign or go unnoticed. Occasionally the mainstreaming of a neologism can offer less than productive framings. We argue that the neologism of big data can fail because it is ambiguous, often deliberately so. In large part the ambiguity derives from a lack of context and intent, which presumably is remedied by small data. Small data can likewise fail to retain intent and neglect the diversity in perspectives among researchers in both their theoretical and methodological understandings of data. In a special *GeoJournal* issue on big data and geography Burns and Thatcher editorialize on the consequence of a neologism that provides less than productive framing: "In organizing this issue, it became clear that even amongst a small set of authors working from a single set of prompts, big data, its influence upon society, and its meaning in day-to-day life will differ radically depending on the research focii contested as important, distinctive, or superfluous. What one author clearly demonstrates as a fundamental concern to epistemology stemming directly from big data analysis, another accepts as a prerequisite for consideration of another fundamental focus."[69]

Sieber and Tenney

Both Big and Small Data Experience Information Loss

Volume is the most important part of the neologism of big data. If size matters in the neologism, then the adage "more is better" captures the homage small data must pay to big data. Kitchin explains that big data lays claim to an exhaustive observation space where entire populations are captured compared to the planned sampling strategies representative of small data.[70] However, capturing entire populations is hardly the case in contemporary big data due to restricted access, unavoidable selection bias, and numerous other factors (e.g., digital divides of potential contributors and differing ontologies of online data sources). This focus on size is perhaps rooted in a conflation of data and information. Wu is a principal data scientist at a big data firm but remains a skeptic. He explicates the "more is better" fallacy: "While data does give you information, the fallacy of big data is that more data doesn't mean you will get *'proportionately'* more information. In fact, the more data you have, the less information you gain as a proportion of the data. That means the information you can extract from any big data asymptotically diminishes as your data volume increases."[71]

Paradoxically the more data one has, the more information one may lose. The signal can become swamped by the noise and the biases. Small data supposedly offers greater contextual comprehension—data geared for the human brain—and therefore could decrease information loss. However, information can be lost in applying the neologism of big data to small data. The previous section mentions how Jacobs detailed the potential information loss to census analyses because the analysis cannot maintain the data's topology.[72] This holds whether the data is big or small (recall that census data are considered small data by some). The analysis space may still be insufficient to the task. If small data is randomly sampled or "analytical-ized" in a fashion similar to big data, then any underlying structure (e.g., the sequence of records) likely will be destroyed.

Big data results in information loss about its life cycle, primarily due to the need for repurposability. However, information loss about the life cycle of data collection can be obscured even with small data. Armstrong and

Armstrong critiqued Statistics Canada's approach to collecting national census data.[73] They recommended reexamining data from the lens of those it was meant to represent and explicating data's relation to the theoretical assumptions made throughout the various stages of each datum's life cycle. This matches a popular ailment of big data in which the context of a datum's creation is as important as the datum itself. As Snickars says in his critique of data mining used by companies like YouTube, "if content is king, then context is its crown."[74]

Small Data Can Lose Verification as Easily as Big Data

Small data like a census seeks to be exhaustive in terms of capturing social demographics on entire populations at set periods in time. Census data lack the velocity and variety to be considered big data; such data are also constrained in terms of access because availability is restricted to sampled profiles. To protect the privacy of citizens, Statistics Canada limits the reporting of certain geodemographic characteristics to a 20 percent sample and provides the geodemographics in aggregated form unless the agency grants special authorization. The advantage of this official or authoritative data set does not necessarily lie in its verified account of the population but in that it is a controlled and directed collection. This is consistent with the domain-based category of small data with its own internally consistent rules with regard to quality control, documentation, and rigor. These rules offer tangible means for understanding possible biases of samples and/or the entire data set, with the potential to compensate for systematic errors.

Elevation of the census as the quintessence of small data implicates this type of data source as a more verifiable or authentic account of social insights than big data. One would be hard-pressed to gain similar levels of geographic or demographic granularity through popular social media services for numerous reasons, whether because of restricted access to proprietary data or inherent degrees of uncertainty induced by unverifiable profile information. Wilson illustrates the unverifiable nature of social media data with a ruse turned viral reporting on the death of actor Morgan Freeman.[75] As Wilson reminds us, truth is no prerequisite of big data, but mistaken authenticity is possible also in small data.[76] The fifth most common religious affiliation reported to the United Kingdom's census was "Jedi Knight." The case of hundreds of thousands of Jedi knights inhabiting the British Isles is perhaps more easily detected than biases found in big data, where teenagers falsely report their ages to circumvent policy restrictions on certain web services. Herein lies a paradox. There are likely more demands on the accuracy of small data than big data because of the former's domain-based architecture of data control. Yet H. Miller and others argue that small data would benefit from the domain-independent analytics and novel data quality methods that were devised for big data. This further impedes, as mentioned above, the goal of using the human brain as the analytical computer rather than depending solely on the algorithm.

Small or Large, Researchers Become Big Data Scholars

An independence of specific domains implies a transformation in scholarship in those domains. Small data research could turn scholars into junior big data scholars. As small data is upscaled, Kitchin sees new opportunities for data science and increased availability of research funding.[77] This mirrors the emergence of GIScience, mentioned earlier. GIScience, emerging from an academic backlash over cultural implications, as well as from a tool-versus-science debate about geographic information systems, centered on whether positioning GIS as a science conferred greater legitimacy to the research.[78] GIS as a form of tool using can be seen as inferior to a GIScience. Tool using represents the domain of practitioners, whereas a science label could lead to greater standing in the academy, with the promise of more highly rated publications, larger grants, and more tenure lines. Transforming big data into a science and then positioning small data within big data could presumably achieve benefits similar to GIScience. We already see the positioning regarding tenure lines with advertisements for academic positions in geospatial data science. If we can aggregate small data sets, for examination with data mining, then small data sets achieve a renewed and rebranded value in the academy.

According to Kitchin and Lauriault, "The data from these studies will more and more be pooled, linked, and scaled through new data infrastructures, with an associated drive to try to harmonize small data with respect to data standards, formats, metadata, and documentation."[79] Exhortations to harmonize data do not automatically result in harmonization, in large part because these digital forms of standardized aggregation can conflict with institutional cultures. Culturally the ideals of the ivory tower may follow the democratic virtues touted by supporters of data sharing. That same culture can punish the pooling of data. The adage to publish or perish remains deeply embedded in research culture. In an increasingly neoliberal university, which injects market values like competition into academe, sharing data by enabling its pooling can mean a researcher loses one additional opportunity for career advancement and job security. Indeed structuration itself can form part of research discovery: "Scientists now have too much choice when it comes to data formats. In fact, it's quite common for researchers to invent formats for each new technique and sometimes each experiment. This makes the work of integrating large data sets significantly more difficult."[80]

Trevor Garrett is lead researcher on the Dutch national project to create an international data-sharing infrastructure.[81] He argues that effective scaling through data structures resembles a kind of magical thinking. Infrastructures may be desired but fail to even approach their asserted objectives. Canada's auditor general disclosed information about a taxpayer-funded $15.7 million project to build a "trusted digital repository for records, but due to a change in approach it was never used."[82] The goal was to collect government data back to 1890, yet the host of the repository, Library and Archives Canada, currently has a backlog of almost one hundred thousand boxes, some of which have been untouched for more than twenty years. The repository's search functions are reported as inefficient, which is particularly problematic with respect to information on Canada's shameful Indian residential school system, which is needed for the Truth and Reconciliation Commission of Canada. Magical thinking has pervaded preparations for archiving paper documents. Canada has yet to craft a strategy to manage the imminent arrival of digital-only documents.

More is known about why individuals refuse to share their data than is known about why they would share. Wallis, Rolando, and Borgman surveyed users of or contributors to a sensor network-sharing platform,

Sieber and Tenney

where participants' greatest concern was information loss in the pooling of data: aggregation on these portals separated data from documentation context that would allow for proper attribution to the original contributors of data.[83] When data sharing did occur, it prevailed in person-to-person interactions and not through impersonal digital infrastructures. The authors confirmed that few institutional incentives exist for rendering data interoperable and then using that shared data. When there is little incentive to share data, it is difficult to envision funding support for an infrastructure to make interoperable the "richness and variance that is likely to exist in . . . slices of the long tail of science and technology research."[84]

Whether the issue is small data or big data, enabling interoperability can demand a profound shift in what is valued in the research process. A drive toward interoperability can move the focus in the means-ends idiom. Instead of using data as a means to generate findings, they become an end unto itself. We have seen this shift in GIS implementation, in which data have long achieved a value separate from the reasons for their generation.[85] One struggle in GIS has been documenting data sufficiently to retain institutional memory about its provenance, classification, and intent. Difficulties in creation and upkeep of spatial metadata have long been known; automation has not markedly improved its collection.[86] There also is the challenge of preparing data in a way that anticipates repurposing of that data for unknown audiences and undetermined usages. Repurposability is a crucial assumption of big data, but compliance can move resources from data use to data preparation. Most researchers and practitioners are not meant to be data producers (i.e., producing data for the sake of data) but data collectors, in situations where data fuel predefined objectives.

The Plight of Very Small and Slow Data

We argue that we will see an impact on hyperlocal and very small and slow-to-achieve-results projects amid an urgency to transition to big data and its accompanying data-driven science. These activities risk being transferred to what we call, for want of a better phrase, very small and slow data studies. We contend that very small and slow data is not necessarily subverted by big data but compelled to complement big data approaches.

This realignment occurs on numerous fronts, including creating expectations of having a "bigness" to one's data set, which then represents the importance of a study, the data, and the rigor of its methods, as well as access to resources (e.g., funding set aside for big data–like studies).

Very small and slow data can be considered part of the process of qualitative social science methods, such as case studies, ethnographic reports, or biographical accounts. These data sets may well be normative, for example, exploring aspects of social justice. Very small and slow data is the size at which much of social science data is collected. The data tend to be highly particularized and require lengthy time periods to collect because they supposedly offer nuanced reflections and deep topological relations, are embedded in historical and anthropological contexts, and, arguably, lie within human comprehension. According to Ballantyne, these types of studies should describe the messiness of what happens on the ground and be distilled into stories by which we explicate the data of our research.[87] Some argue that big data allows us to escape an era of scarce data so we can live in data-rich environments.[88] Another perspective is the promise that our meager data stores can describe rich environments. The appeal of these studies is that value can be found in the very noise that gets discarded from big data to achieve the signal. Very small and slow data can be the reasoning or speculation that occurs behind the key-value coding in many content analysis approaches, or it can be the thought processes of researchers and their subjects in determining their "choice as to what is most real."[89] While we are supportive of a very small and slow approach to data collection, even when done digitally, we do not automatically advocate that the only good data would be the smallest and slowest. Numerous reasons preclude these types of studies (e.g., objectives of study, resource constraints, or objections of participants). Our definition for very small and slow data is imperfect and subject to the very critiques we offered above. We choose the term as a provocation and simply question the drive to subsume all data to the assumptions embedded in a neologism.[90]

We rely on the first equation, *big data = f(volume, variety, velocity, and maybe veracity, viability, variability, value . . .)*, and the *V*s to suggest some ways in which this shift is particularly toxic to very small and slow data.

Sieber and Tenney

Normative Positioning through Size

Very small and slow data brings the assumptions of volume into high relief in how big data normatively positions small-scale social science research. Because volume can be measured numerically and categorically (i.e., ordinally), it embeds a hierarchy. Bigger is better. Any hierarchical system or dichotomous pairing presupposes an ethic, either where one choice is instrumentally superior to another choice or where one "ought" to select one choice over another (i.e., there are right and wrong practices). Smallness also reflects the type of data represented. Referring back to the introductory anecdote about Web 2.0 for community development, a small number of VGI observations could be considered flawed relative to a large number, in part because the observations are asserted and not emergent from experts. By implication very small and slow data sets would require strengthening, whether by imposing accuracy or, in a Wikipedia crowdsourced model, by precision. Only by this layering, this accretion of assertions, does one approach value. There may be no refinement, yet the quantity constructs validity. In an epistemology of big data in which ways of knowing are attached to large numbers of contributions that serve to triangulate each other, precision presumably fixes the mistakes.

A larger assumption concerns the way the size of big data (or small data as envisioned with very large data sets like a country's census) convinces us that with volume, reach, and scalability we can attain new insights. In comparison to big and small data, very small and slow data, unless we pool it, limits our ability to maximize insights. Hardt exposes big data biases in his article, "How Big Data Is Unfair." Hardt charts the methods by which big data can dilute minority views, which are statistically overwhelmed by the volume of majority opinions. Instead of reaching the long tail of public opinion, big data can result in a regression to a "white" mean that diffuses minority voices while giving the appearance that minority voices are heard.[91] This relates to Elwood and Leszczynski's observation that we too often conflate availability with access.[92] Just because anyone can participate on a social media platform does not mean everyone will participate. Contrary to the assumption that big data is neutral while very small and slow data is biased, both big data and the analytics offer a social mirror to our biases:

> As we're on the cusp of using machine learning for rendering basically all kinds of consequential decisions about human beings in domains such as education, employment, advertising, health care and policing, it is important to understand why *machine learning is not, by default, fair or just in any meaningful way.*
>
> This runs counter to the widespread misbelief that algorithmic decisions tend to be fair, because, y'know, math is about equations and not skin color.[93]

Speed over Nuance

Small data, as compared to big data, suffers from a lack in "real-time" velocity in both its creation and collection. Very small and slow data can allow us to contemplate the good things that come to those who wait. Often the waiting occurs whether researchers want it or not. An important feature characterizing the collection of very small and slow data is the building of trust between the researcher and research subjects (not to be confused with algorithmically calculated trust used in many big data social network projects). To gather very small and slow data from in-depth interviews, one must allow time to cultivate a personal relationship and trust. Interview respondent numbers may similarly be small in size with perhaps a few dozen respondents; interviews may have no or irregular periodicity (one-time interviews or a sequence over a number of months), possess weak relationality, and be limited in variety (e.g., only text transcripts). Instead of assuming that very small and slow data is weak in insights, the information (insights) gathered from these kinds of data may be richly textured and supported by rigorous methods. Simultaneously, insights derived from slowness of certain methods may be incompatible with an instant-access age, in which data are constantly updated in a continuous stream. Numerous situations may require instant access. We may save lives because of the speediness afforded by citizens' sensing of crises.[94]

Sieber and Tenney

However, in conforming to the assumptions of big data we risk abandoning the slow study in preference to the speedy superficial.

Harmonizing the Smallest and Slowest Data

Finally, let us consider variety. Efforts needed to maintain the value of very small and slow data in a big data future may oblige researchers to ensure their data are linkable and scalable, as in the case of small data. The assumption underlying harmonization is that data gains value in its aggregation. The converse could be characterized as "a pixel unused is a pixel wasted." If the data exist only in the proprietary silo of a research report, then they fail to achieve their potential. Why should such data not be used again? As sharing and reuse, particularly digital repurposing, has become intrinsic to current research, the questions become *argumentum ad hominem*. What does the objector have against the reuse of data, especially if that reuse generates new knowledge? The subtext is that the researcher is immoral for not attempting to wring more insights out of preexisting studies, if insights can be accrued in combination with other data sets. A clear expression of morality lies in the attribution of life-saving properties to linked data: "Examples of the power of linked data arise daily. In Britain, *The Times* picked up raw, linked data about bicycle accidents from DirectGov and published a mashup map showing where bicycle accidents had occurred, so cyclists could be aware of the many dangerous spots along the city's roads."[95]

Very small and slow data therefore resembles this expression:

big data = n*(very small and slow data)

where *n* is the linkage threshold at which the data become legitimate in the epistemology of big data. However, very small and slow data allows us to examine the converse: "why shouldn't we waste the pixel?" Certain data cannot and should not be repurposed. Sacred data exemplifies the conflicted nature of data sharing. Rundstrom wrote that in many indigenous cultures certain knowledge could be known only by small numbers of people (e.g., elders).[96] Others had no rights to that knowledge.

Certain bands would accept the loss of indigenous knowledge if there was no incoming elder rather than allow that knowledge to be recorded. In another instance an indigenous group would allow its sacred site to be destroyed rather than permit it to be mapped and potentially expose that knowledge to a broader public. The supposition that some data sets should be lost and not be repurposed or ever made public violates the ethos that all data should be available for linking.

When we conduct research with or about marginalized populations in very small and slow data studies, we frequently place our research in a critical context. These include positionality and subjectivity vis-à-vis the individuals we conduct research with or on (e.g., "One author is a white middle-class cis-gendered woman co-conducting research with indigenous peoples, who are actually a subset of a larger indigenous grouping"). Technically a harmonized, linked-data approach can attach these details to the extracted data because of their polymorphism (e.g., a document file to an individual record). An initial linking, however, cannot guarantee the link is subsequently maintained. A linking also may exclude an ethics review. Indeed ethics may not permit a repurposing. Whether we are linking or pooling, we could lose much in the harmonization of very small and slow data. Perhaps certain data sets should not scale.

Conclusion

In this chapter we moved from big data to small data to very small and slow data and back again. This structure allows us to meditate on the seesaw rhetoric of big data. Namely, big data is too big or fast to comprehend or to manage computationally. It fails to produce value as advertised, so we shrink the data to a manageable size.[97] Small and very small and slow data offer value through purpose-driven data collection, but they can be considered inconsequential for newer analytics, visualization, and, ultimately, insights. Consequently we are urged to employ various aggregations to scale them to big data. But the resultant data may lose their context and become too big to comprehend. So we shrink the data; repeat the rhetoric as needed.

We do not argue against the value of any one size (i.e., the *V*s suggested by big data) of data set over another. Instead we argue that the hyperbole of big data permeates all data. Regardless of data size the temptation is to position all data within the opportunities—the insights and the new valuations—offered by big data. Generalizing one size of data as being representative of all social science research misrepresents the nature of small and smaller data and the value of all sizes of data in the future of a discipline such as geography. Even considering small data as a unary representation of geography issues a misapplied philosophical reduction to the discipline. Traditional unary representations and neologisms serve more to artificially obfuscate our work than elucidate the discipline's future. We hoped to demonstrate this by the ironic coinage of our own term: very small and slow data.

We expect more numerous calls for very small and slow data to be repositioned as amenable to small data, which in conjunction is repositioned as a contributor to big data. We prefer the acceptance of diverse data sets and of varied approaches, one of which has some data never warehoused, shared, or linked. In this chapter we sought to deflate some big data hubris being prematurely attached to small and smaller data and which likely will face difficulty in matching the exaggerated claims regarding the utility of resembling "big." We urge, along with many others, restraint in adopting such epistemologies for social science disciplines like geography, because they miss bigger issues that could derail the relevancy of big data in the future of social science research. By adopting the pluralistic acceptance of different sizes of data in geography, we should strive to balance the critical and the opportunistic somewhat in the fashion of M. Graham and Shelton: "We believe that a broader conversation into the big data meme itself and the ways that it is able to redirect and displace attention, conversation, resources, and practices away from other pressing issues will not only allow us to avoid the most problematic implications of big data but also work toward a more productive integration of big data with existing research paradigms."[98]

Notes

1. Ruppert, "Rethinking Empirical Social Sciences."
2. Simon, *Too Big to Ignore*.
3. Schmidt and Cohen, *New Digital Age*.
4. Buytendijk, *Hype Cycle*.
5. See Lerman, "Big Data"; Kaisler et al., "Big Data"; and Lazer et al., "Parable of Google Flu."
6. Wilson, "Morgan Freeman Is Dead," 345.
7. E. Sheppard, "Quantitative Geography."
8. Wilson, "Morgan Freeman Is Dead."
9. Hidalgo, "Saving Big Data."
10. Mashey, "Big Data."
11. Mashey, "Big Data."
12. Weiss and Indurkhya, *Predictive Data Mining*.
13. Weiss and Indurkhya, *Predictive Data Mining*, as quoted in Diebold, "On the Origin(s)."
14. Hey, Tansley, and Tolle, *Fourth Paradigm*, 1.
15. Elwood, Goodchild, and Sui, "Prospects for VGI Research," 371.
16. Anderson, "End of Theory."
17. Anderson, "End of Theory"; Bar-Yam, "Limits of Phenomenology."
18. Hey, Tansley, and Tolle, *Fourth Paradigm*, 223–24.
19. Elwood, Goodchild, and Sui, "Prospects for VGI Research," 371.
20. Raymond, *Cathedral and Bazaar*.
21. Laney, 3D *Data Management*.
22. Dutcher, "What Is Big Data?"
23. Floridi, "Big Data."
24. M. Graham and Shelton, "Geography and the Future of Big Data," 256.
25. Batty, "Big Data, Smart Cities and City Planning."
26. M. Chen, Mao, and Liu, "Big Data: A Survey."
27. Wilson, "Morgan Freeman Is Dead."
28. Nelson, "Big Data."
29. Diebold, "On the Origin(s)."
30. Baldwin, "Big Data as Boogeyman"; Peters, "Big Data Gold Rush"; Press, "6 Predictions."
31. Davenport, Barth, and Bean, "How 'Big Data' Is Different."
32. Floridi, "Big Data."
33. Infochimps, "CIOs & Big Data."

34. Greenberg, "10 Reasons."

35. Fidelman, "These Smart, Social Apps"; Walker, "Small Data Is Beautiful."

36. Walker, "Small Data Is Beautiful."

37. Bonde, "Defining Small Data."

38. Kitchin and Lauriault, "Small Data," 464.

39. See, e.g., Barber and Harfoush, "Synchonizing Small Data"; daCosta, *Rethinking the Internet of Things*; and A. Paul and Bruns, "Usability of Small Crisis Data Sets."

40. M. Chen, Mao, and Liu, "Big Data: A Survey."

41. Davies and Frank, "There's No Such Thing as Raw Data."

42. J. Lu and Li, "Bias Correction," 2658.

43. Jacobs, "Pathologies of Big Data."

44. Jacobs, "Pathologies of Big Data."

45. Jacobs, "Pathologies of Big Data," 40.

46. Floridi, "Big Data," 436.

47. Buhl et al., "Big Data."

48. See, e.g., Gutierrez, "Big Data vs. Small Data"; Pollock, "Forget Big Data"; and Markowsky, "In Praise of Small Data."

49. Markowsky, "In Praise of Small Data."

50. Markowsky, "In Praise of Small Data," 1.

51. Couldry and Powell, "Big Data from the Bottom Up."

52. Rast, "Context as Assumptions."

53. Simon, *Too Big to Ignore*.

54. Small Data Lab at Cornell University, http://smalldata.tech.cornell.edu/; Estrin, "Small Data," 33.

55. Elwood and Leszczynski, "Privacy, Reconsidered"; Estrin, "Small Data."

56. Simon, *Too Big to Ignore*.

57. The special issue was *GeoJournal* 80, no. 4 (2015).

58. Goodchild, "Quality of Big (Geo)Data." Whereas we have provided several distinctions of small from big data, authors and advocates tend to emphasize only the first *V*, volume. If governments are prime producers of small data, then we certainly see considerable variety in small data, for example, data from water mains, transportation, and park locations. Additionally censuses satisfy a velocity requirement of big data. Censuses have a temporal periodicity, for example, yearly or decennially. Small data allows us to refine the characterization of velocity to high and increasing speed. The velocity component of big data would more accurately be called acceleration.

59. Kitchin, "Big Data and Human Geography," 3.

60. H. Miller and Goodchild, "Data-Driven Geography."

61. See Couclelis, "Construction of the Digital City"; Sui and Bednarz, "Message Is the Medium"; and E. Sheppard et al., "Geographies of the Information Society."
62. Goodchild, "Quality of Big (Geo)Data," 8.
63. Kitchin, "Big Data and Human Geography," 263.
64. Hyman, "Small Data."
65. Goodchild, "Quality of Big (Geo)Data."
66. Kitchin and Lauriault, "Small Data," 463.
67. Kitchin and Lauriault, "Small Data," 464.
68. Haklay, Singleton, and Parker, "Web Mapping 2.0."
69. Burns and Thatcher, "Guest Editorial," 3.
70. Kitchin, "Big Data and Human Geography."
71. Wu, "Big Data Fallacy."
72. Jacobs, "Pathologies of Big Data."
73. Armstrong and Armstrong, "Beyond Numbers."
74. Snickars, "If Content Is King."
75. Wilson, "Morgan Freeman Is Dead."
76. Wilson, "Morgan Freeman Is Dead."
77. Kitchin, "Big Data and Human Geography."
78. Wright, Goodchild, and Proctor, "Demystifying the Persistent Ambiguity of GIS."
79. Kitchin and Lauriault, "Small Data," 464.
80. James, "Out of the Box," 119.
81. http://www.etriks.org/author/trevor-garrett/.
82. Office of the Auditor General of Canada, "Chapter 7."
83. Wallis, Rolando, and Borgman, "If We Share Data."
84. Wallis, Rolando, and Borgman, "If We Share Data," 15.
85. Onsrud and Pinto, "Diffusion of Geographic Information."
86. Olfat et al., "Spatial Metadata Automation."
87. Ballantyne, "Geomatics and the Law."
88. H. Miller and Goodchild, "Data-Driven Geography."
89. Mills, *Sociological Imagination*, 67.
90. Smallness and very small and slow are subjective classifications. Ethnographers would not likely qualify their work as "very small." A census worker would hardly consider a census small, even with the definition in Kitchin and Lauriault, "Small Data." We acknowledge these disagreements even as we use the terms.
91. Hardt, "How Big Data Is Unfair."
92. Elwood and Leszczynski, "New Spatial Media."
93. Hardt, "How Big Data Is Unfair" (original emphasis).

94. Goodchild, "Citizens as Sensors."

95. Fischetti, "Web Turns 20."

96. Rundstrom, "Teaching American Indian Geographies."

97. Floridi, "Big Data."

98. Graham and Shelton, "Geography and the Future of Big Data," 259.

4   Reflexivity, Positionality, and Rigor in the Context of Big Data Research

Britta Ricker

In this robust unstructured era of big data, neopositivistic empiricism asserts strict objectivity when manipulating data, yet big data is riddled with the subjective positions of those entering the data, those creating and maintaining the storage and retrieval mechanisms, and those sifting through the data. Big data offers unique analytical opportunities to reveal patterns that may otherwise have gone unnoticed. The sheer amount and variety of data being generated by a seemingly heterogeneous population and then collected primarily by businesses constitutes a relatively new phenomenon, and it is tempting to assume data do not lie and are truths.[1] The experiences and subject positions (the positionality) of those entering the data are often missing or lost when amalgamated into big data, and the epistemologies of those who manipulate and handle the data come to inform the methods chosen to shape the meanings drawn (or not) from the data entered. The positionality (in terms of race, age, socioeconomic status, ethnicity) of a researcher, scientist, database administrator, and other actors influences what questions are and are not asked in data science. Knowledge is mediated and constructed through interactions with the world. The meanings extrapolated, the knowledge built from big data, are limited by the questions that are asked.

Here I impose a social constructivist critique or rather a reflectivist theoretical stance.[2] Echoing the postulation of boyd and Crawford and of Dalton and Thatcher, I infer that the questions we ask, the analysis we conduct, and in turn, the patterns we find in big data are heavily influenced by our epistemologies.[3] Epistemologies describe the background and the perspective or lenses through which we study the world.[4] We must consider not only the perspectives of individual researchers but also those of other individuals, such as the computer scientists and database administrators who maintain the data.[5] The so-called black box is made up not of a conspiring individual but of a disconnected stream of bureaucracies, or countless groups of individuals limited by their own knowledge, positionality, time, and other constraints, including those associated with actors upstream. Thus I ask, (How) do identities and experience (dis)appear from big data? This overarching question can help explain research design choices and what meaning is extracted from big data based on the identities and experiences of the researchers who enter the data and the database administrators who maintain the data. This can also be applied to the identities and experience of the subjects whose data have been amalgamated into big data. As critical theorists moving past positivistic assumptions, we understand that even the choice of mathematical methods by which researchers query, analyze, and display data is influenced by their positionality. While big data may seem to be overtly quantitative, the data are also overwhelmingly qualitative in nature, necessitating methodology distinctive to qualitative research.

In an effort to identify social and spatial implications hidden (or not) within big data, we can learn lessons from qualitative geographic information systems (GIS) and previous GIS critique that can provide useful insight when investigating ways in which to productively utilize big data, particularly for social benefit beyond existing neoliberal initiatives.[6] In the tradition set by Baxter and Eyles, England, and others who call for a high standard of rigor in qualitative methods, I suggest that we enter into a discussion considering how these same standards of rigor applied to qualitative research could be beneficial for the analysis and manipulation of big data.[7] Theories and reflexive methodologies associated with feminist geography are becoming richly diverse; they should be creatively applied to the study of big data.[8] Otherwise we risk continuing to serve the agendas of the technocratic elite, rather than the needs of the underprivileged— those who could benefit most from sharing their personal information.

Big Data in a Spatial Context

Big data is considered to be data that are high in volume, velocity, and variety, as well as highly flexible and exhaustive and often containing associated locational and temporal information.[9] Big data is seen as perplexing due to the sheer amount of data and because of the recurring and unspecified filtering imposed on them.[10] Data that previously went (digitally) undocumented are now being collected via the big data movement. These data are frequently diverse, consisting of, for example, time stamps, documenting not only such things as a phone user's most recent activity, movement patterns, buying habits, and call logs but also a wide range of ambiently collected data. Other data being collected are qualitative, full of rich information such as opinions typed into a restaurant review application. All of these activities can be and often are documented by someone and something. While these data are new in that they are being collected digitally in volumes that were once considered infeasible, the idea of collecting this type of data is not entirely new. Mathematicians of the past have encountered some of the same challenges being faced today by big data engineers.[11] This form of digital data offers social scientists, particularly social spatial scientists, new opportunities to collect and access qualitative data through the utilization of massive amounts of unstructured data.[12] While excessive amounts of digital information are being collected and hoarded at present, the goal of harnessing their full potential has yet to be realized. Utilizing big data is a significant challenge because the constituent data are for the most part unwieldy. Businesses and researchers alike are still figuring out how to interact with this form and volume of data. Big data has the potential to reveal and link spatial contexts, meaning, and processes, but these data are noisy and cumbersome, making it difficult to yoke their strengths for social good.[13] Kitchin clearly identifies opportunities, challenges, and risks associated with big data and spatial science research. Opportunities are afforded in the sheer amount and diversity of data being collected, while the risks and challenges associated include the assumptions that these data can "speak for themselves" and that theory is dead.[14]

The data entry process associated with big data may inform experience in place. These are the contexts in which the data are collected. They are often the location of the qualitative observations most valuable to human geographers and social scientists. The data being entered and then aggregated as location-based services (LBS) amounts to a cyclical process.[15] Early research suggests that many voices and experiences (particularly those of the underprivileged and underserved) are missing from big data currently being collected.[16] Big data collection risks infringing on individuals' privacy while also exhibiting social unevenness.[17] Experience from the past could help inform how big data can be used for productive purposes.

While the focus of this chapter is not on the makeup of big data infrastructure, I believe it is relevant to point out that there are multiple levels of data collection, storage, aggregation, and manipulation, thus introducing many opportunities to insert personal bias in one stack. Broadly speaking, the stack refers to the combination of elements, typically software and hardware components, required for a database or system to be usable and useful. If you were to search online for images related to the term "big data stack" you would find thousands of versions of different data structures. The configuration of the stack dictates what information can be collected and how it is combined or not with other data, and each of these ostensibly minute decisions about collecting and combining influences how data can be used in the future. It is these seemingly objective tasks and steps where it is possible for those who interact with the data and make decisions about the receptacles and organizational structures in which it is housed to insert bias. Descriptions of these stacks include a storage location, often in distributed data centers, file systems, data warehouses, and databases. There are endless ways to configure a big data stack. Like research more broadly, the process of analyzing and organizing this stack associated with (big) data is recursive and, ideally, reflexive. Technical and theoretical research is being conducted to identify relationships between datacentric and operation-centric approaches to computationally intensive geographic data and analysis methods.[18]

To retrieve data collected and stored, a search needs to be conducted using a query language. Conducting such a query across distributed databases and data centers is no easy task. This complex process is constantly being improved, and much research is being done to optimize these types of workflows specialized for big data handling with a special consideration on spatial information.[19] Once data are retrieved they must be aggregated. There is no single industry-standard software package available at this time that can retrieve the data, run statistical analysis, and provide visualization capabilities, since at present there are three or more software systems required to fulfill all of these tasks. I consider the software package de jour for data retrieval to be Oracle NoSQL, for statistics it would be R, and for visualization the popular choice in 2016 is Tableau (depending on the circle). At present Hadoop is a popular open-source big data software framework for setting up hardware clusters. The workflow vaguely described here varies dramatically from project to project and company to company. This is an evolving process that influences who handles data and who in turn may modify it. Data provenance is a formal area of research.[20]

When big data are described as passing through the black box of software and data aggregation, it is important to understand that the black box can be thought of as a bureaucracy in that many actors are involved in setting up and maintaining "the stack." The stack refers to the organizational and flow structure of the data, where data are stored, the direction in which data travel, and where data are housed until called upon. Each of these actors has its own positionality, educational background, and understanding of the technology that is available and accessible. These perceptions include views and experience with proprietary versus open-source software systems, which will in turn affect the storage, access, retrieval, and visualization of information down the line. Each of these actors has different time constraints, which will also influence decisions about how to set up and maintain a big data stack. This is the description of just one of many other potential stacks. As Thatcher points out, the "data fumes" from one system could be passed along to another.[21] Taking this view a step further, I suggest that the organization of a stack could influence data that are included and how they pass through several other stacks before they reach an end user or researcher, leaving endless opportunities to change the data.

The technical architects associated with a project largely dictate big data stack construction and decision making, which influences the structure

Fault–tolerance

FIG. 4-1. Example of how power relations manifest themselves in the workplace; knowledge of a system influences ability to access and manipulate data. Illustration courtesy of John Muellerleile (@jrecursive).

and maintenance of a data stack. These decisions are influenced by the goals of the individual project, the funding agency, and the experience of those on the development team. These decisions have long-term ramifications, including if and how spatial data are collected, stored, and used. Each actor associated with the stack holds power, as depicted in a popular Internet comic (fig. 4-1). In this comic one colleague is asking another to pull data from a database. It becomes clear, based on the terminology being used in the exchange, that the person who is asking to pull data from the database does not know how to do so from a big data stack. It can also be inferred that the person being asked does not want to complete the task and is thus engaging in a display of power.

In an effort to demystify big data for researchers approaching it from the geoweb, Thatcher's research revealed the actors who control the black box by illuminating the positionality and the thought process of those creating the structure of the data—not only the data collected or the output but also the way in which the data are housed and queried as well.[22] His work has provided a fundamental understanding of the decision-making processes and structure associated with big data collection and distribution.[23] Dalton and Thatcher have provided glimpses of what is inside the black box.[24] Their efforts can be used to inform future efforts in utilizing big data for social improvement.

Lessons from Qualitative Research and Rigor in the Big Data Context

While qualitative geographers and social scientists call attention to how their work cannot and should not be evaluated in the same manner as positivist/quantitative research, they have rarely described how their research *should* be evaluated.[25] Feminist poststructuralists, challenging the so-called objective research paradigm, call for ethical research by stimulating researchers to be more reflexive and inclusive of methods that reveal sensitive power relations between the researcher and those being studied.[26] As researchers of big data, we must locate ourselves in our work and show how this location influences the questions we ask and how we conduct and write up our research.[27] We need to identify ourselves as insiders or outsiders in relation to the subject in an effort to help us interpret the meaning of the data collected.[28] How much are we participating in the big data process and at what point in the process do we as researchers have experience?

Human and feminist geographers challenge us to be suspicious of objectivity and move beyond looking only at spatial patterns by instead looking at other types of relationships within our research.[29] This is not to say that qualitative methods are not scientific or rigorous. In quantitative research the term "methodological rigor" refers to the validity, reliability, and objectivity of the research process. Within the field of qualitative research this includes researchers' responsibility to be reflexive and honest about their position in relation to the research being conducted.[30] Rigor in both qualitative and quantitative research helps indicate at what level the research is believable and worthy of attention.

To establish rigor in qualitative research, Baxter and Eyles suggest a set of criteria rather than a fixed set of rules. Their suggested criteria include the credibility, transferability, dependability, and confirmability of one's research. While these authors were evaluating and speaking largely to qualitative research in which the researcher had direct contact with the subject when collecting data through interviews and other forms of direct contact with a community, these ideas can and should be applied to the study of big data. Baxter and Eyles conducted an extensive literature review in which they identified specific conditions as indicators of rigor in research: the presence or absence of a rationale for methodological choices, multiple methods, description of respondents, direct quotations from subjects, interview practices, procedures for analysis, immersion and length of fieldwork, revisits, verification by subject, appeals to interpretive community, and rationale for verification.[31]

Debates from GIS Critique as They Apply to Big Data

The aim of a critical GIS is to illuminate ways in which positivist practices associated with GIS and knowledge production generate hierarchies of power that produce social, economic, and cultural inequity.[32] Critical GIS and qualitative GIS scholarship has extensively covered the debates associated with the critique of GIS research, questioning quantitative versus qualitative approaches to the use of the technology and extending this critique to the geoweb.[33] It is not my intention here to provide an extensive literature review related to these ideas or debates but rather to echo M. Graham and Shelton's call to look to experiences in related disciplines to highlight ideas and arguments that could be relevant to the critique of big data in a spatial research context.[34] Debates that occurred primarily in the 1990s regarding epistemological approaches to research in GIS are well documented yet still unresolved.[35]

We risk replicating epistemological debates documented in critical GIS literature in the context of big data and the geoweb, because like GIS, big data is seen as positivistic and objective. Quantitative ways of interpreting data and representing those data as maps that are associated and produced with GIS are given more weight and seen as authoritative compared to other representations of knowledge. GIS has been critiqued particularly for affording scientists a disembodied view of the world, often referred to as the "god trick," referring to the ability to see the whole world while being distant from it.[36] Historically GIS and cartography have not commonly featured minority subject positions or objects, and thus the aim of the critical GIS movement has been to eradicate socially constructed inequalities through maps.[37] Significant efforts have been increasingly utilizing GIS for feminist endeavors.[38] Recent research harnessing big data on the geoweb suggests that women and underprivileged populations' views are not being equally represented using traditional forms of GIS.[39] Similar debates and findings are likely to arise in the context of big data and the geoweb.

Referring to the data being collected, GIS are socially produced, creating a case that can be researched using multiple epistemologies.[40] As Elwood and Cope postulate, it is valuable to utilize multiple episte-mologies, through diverse modes of analysis and forms of knowledge, to transcend levels of agency and authority through different forms of data, representation, and analysis in research. Epistemology is closely intertwined and influenced by positionality.[41] Positionality influences epistemologies, as well as the data collected and consequently the infor-mation that becomes knowledge generated therefrom.[42] Critical reflexive engagement challenges quantitative and positivist approaches closely held by traditional research in GIS.[43] Knowledge is situated and not objective.[44] Feminist scholars acknowledge that situated knowledges add diversity to our understanding of the world, thus providing an opportunity to engage in dialogue with those possessing other positionalities.[45] Recognizing that knowledge is situated requires researchers to reveal their positionality, which in turn reveals the origin of the truth being documented, thus giving the researcher ownership of and responsibility for that truth. This process has been termed "strong objectivity," acknowledging that positionality influences worldview, and recognizing this is a step toward objectivity in that it proclaims that neutrality in science is impossible.[46] It has also been recognized that units of knowledge are not fixed; they are fluid and evolve and can pool to create shared bodies of knowledge and need not be considered distinct experiences that influence who, whether groups or individuals, counts in a GIS.[47]

With situated knowledge in mind, "GIS scholars work with a mixed epistemological toolkit that varies from positivism to pragmatism."[48] It is necessary to acknowledge that methods and epistemology can be related but are not fixed; multiple epistemologies can utilize a single method.[49]

Critical cartography has taken steps to acknowledge and celebrate inclusive mapping practices that are occurring outside of corporate and government agencies, which typically are responsible for managing spatial data and cartography.[50] By advancing non-normative claims to power, one can advocate for those who are typically marginalized by bringing attention to seemingly inevitable dominant positionalities typically associated with power.[51]

At present big data is most commonly collected and analyzed to push corporate agendas. These agendas have been masked and defended by claims that data are objective. Quantitative ways of presenting information are given more value than qualitative information, which is often equated with anecdotes. When GIS was originally gaining popularity within the field of geography, it was critiqued for serving elitist technological agendas.[52] Those critiques are similar to the way big data is now being critiqued in this book. Feminist poststructural critiques recognize that multiple perspectives and methodologies can be inserted into GIS.[53] Similar observations and transitions will likely be observed for big data, as more and more of the subject positions of those involved in the process of knowledge production resulting from big data are revealed. Much like GIS, we risk reproducing in our big data the inequalities observed in reality on maps.[54] The medium used to collect, analyze, and aggregate data will influence who participates in the process of knowledge production.[55] It is hoped that as more people identify the utility in big data and the skills required to harness the utility, more will participate in the analysis thereof, making more diverse epistemological representations possible in the future.

Inserting Reflexivity through Positionality in the Structure of Big Data

Bringing the discussion back to the promise of big data, M. Graham and Shelton call for use of big data to identify ways to reduce social inequality and environmental injustice.[56] A critical first step is to begin rigorous qualitative research, including data collection, analysis, and dissemination associated with the big data utilization process.

Wilson asks who gazes into the (Arc)toolbox and how. The same question can be posed in terms of big data: who can see and reach into the black box?[57] As researchers we need to explicitly state our positionality. Are we insiders or outsiders in our relationship with the technology in question? Wilson clearly describes this spectrum as a borderland in terms of GIS researchers—the insiders who practice GIS and the outsiders who do not.[58] When we consider this idea in terms of big data, what is missing are the mechanisms associated with collecting, storing, retrieving, and analyzing the (big) data. While big data is ostensibly a technological challenge, technology is never utilized outside of social constructs. Thus it is necessary to discuss epistemological approaches to technological utilization associated with big data.[59] We must consider using these new processes associated with qualitative data collection and new methods associated with big data so we can work toward a more inclusive big data horizon.

Inserting reflexivity into our own research, but also considering the positionality of those who are invisible actors in the supply chain of big data, may insert rigor into our research in big data. The aim of reflexivity is to remind researchers that they are not machines. Context and subjective perspectives shape the meanings the subject shares with the researcher.[60] Here it is important to remember that individuals, who maintain epistemological individualism, program the machines.

Big data is made up of seemingly invisible processes, yet it is possible to visualize the invisible, to apply qualitative rigor by considering the positionality associated with those who are programming the cyborg.[61] While it is not possible to connect with each of the individuals involved in collecting, storing, and retrieving big data, it is possible to contemplate their positionality. Consider not only those who input the data and then analyze the output but also those who create and construct the data infrastructure in which the data are stored. They may inadvertently provide an opportunity to insert bias, to reflect the positionality of the developer, the data infrastructure engineer, and the cyber construction worker. The position of the researcher and the programmer (and those maintaining big data) may influence the accessibility of information; others may not have access.[62] These individuals controlling the data have unique constraints, including time, money, expertise, and previous experience (including epistemology). Other limitations may include what can and cannot be collected in each data field and what data can be collected by a sensor or by a query. For research to be replicable the positionality of those involved in the data collection, storage, and analysis procedures needs to be revealed or at least considered, because those with different positionality may collect different data or seek patterns in data that have been hoarded. Also invisible is the positionality of programmers. What technical training have they received? Why did they choose open source versus proprietary software?

Discussion

In the tradition of Baxter and Eyles, England, and others who have set the standard for rigor and qualitative methods, I suggest that we enter into a discussion regarding how to apply qualitative rigor when analyzing, manipulating, and critiquing big data.[63] There is a need to be reflexive as researchers while we develop and pose our research questions and assert our positionality as it applies to both the collection and analysis of big data.

Inserting qualitative rigor into a quantitative field not only introduces reflexivity but also reifies the inclusion of more voices in the data set, meeting the goal of using big data for social good. Baxter and Eyles have suggested criteria to establish rigor in qualitative research in a different context.[64] I suggest that we find ways to apply such criteria to the analysis of big data. These criteria include consideration of the credibility, transferability, dependability, and confirmability of one's research. In addition to the consideration of methodological rigor itself, particularly making explicit the rationale for methodological choices, indicators for rigor in research include the use of multiple methods, a description of respondents, direct quotations from subjects, interview practices, procedures for analysis, immersion and length of fieldwork, revisits, verification by subject, appeals to the interpretive community, and a rationale for verification.[65]

It is clear that researchers investigating social space see much hope in using big data to help underprivileged populations; however, we have yet to see this hope realized.[66] These scholars are echoing England's recommendation to conduct research with integrity, especially among marginalized populations. Feminist poststructuralists are calling for ethical research, thus challenging the so-called objective research paradigm by appealing to researchers to be more reflexive and to use methods that reveal sensitive power relations between the researcher and those being researched.[67]

While the process of interacting with big data is opaque and the researcher is commonly far removed from those entering the data or even organizing and storing it, qualitative rigor could be incorporated by simply recording these positionalities in the comment section of code, by adding it to the metadata, or by considering the positionality of all of the actors involved in the process, that is, the positionality of not only the researcher but also those who control other points in the big data process.

Even if we overtly acknowledge that the research relationship, including the big data stack, is hierarchical, reflexivity alone cannot resolve the issue, but it will make others more aware.[68] While the researcher holds power (in the case of big data, the powers that own the data or those with the skills to interact with the data), she can also give it away.[69] There is a need to be reflexive as researchers while we develop and pose our research questions, as well as to assert our positionality as they apply to both the collection and analysis of big data.

Acknowledging the multiple positionalities involved in assembling big data used in qualitative research may help the researcher gain access to new information. The same consideration can be applied to those building the infrastructure for big data.[70] Those who pose the query to the key-value store pair may (not) gain access to information based on their positionality, their understanding of the world, and what they can offer the data contributor in return.

It is the researcher who decides which voices will be heard and what will be written up in the final project to be shared with a wider audience; reflexivity and acknowledging the researcher's position in the research constitute an example of rigor.[71] In the era of big data there are so many actors involved that they all have had some hand in determining what data are included in a data set that may then be analyzed by a researcher or analyst. No longer is it only the researcher who decides which voices are heard, which data are collected, and which data are queried, stored, and aggregated.[72] Today big data is stored within large bureaucracies. To be considered, individual bits of data must make it through several steps before they reach a researcher. No data are truly raw.[73]

Conclusion

I have described here, from a qualitative perspective, some ways in which feminist and poststructuralist approaches could be applied to research challenges that are associated with big data research. In the future, by identifying ways to use big data to reveal (spatial) patterns and qualitative experiences that previously went unnoticed, established qualitative research methods could be used to impose rigor and to identify valuable patterns within the big data. How big data is interpreted is informed by individual experience with it in the world and by epistemologies of everyone who touches it. It is hoped that big data utilization can go beyond number crunching, toward the discovery and display of spatial patterns of behaviors or needs; however, we must acknowledge that the mathematical methods by which we choose to analyze and display data are influenced by our positionality. Guidance offered by Baxter and Eyles, England, and others has informed us of the utility of rigor in qualitative research methods, and here I have displayed how these ideas can be used to investigate big data.[74]

Attempting to illuminate individual positionalities within the black box or the stack could be considered an effort to mitigate positivistic assumptions and call into question how subjectivities are inserted in the collection, storage, retrieval, and analysis phases of big data, thus inserting rigor. Facilitators of each step within a stack have an opportunity to insert their subjective bias; their positionality influences the data. By acknowledging the positionalities of those working within the stack, we come one step closer to "strong objectivity." Understanding this relationship will help inform how we approach big data as qualitative researchers. The data cannot speak for themselves, as they have passed through many hands, many boxes, and many fields prior to reaching the hands of the researcher or analyst.[75] The black box contains these stacks made up of disconnected streams of bureaucracies, comprising not necessarily conspiring individuals but a group of individuals limited by their own knowledge, positionality, and time constraints.

Much can be learned from scholars in other disciplines who critique similar technological approaches to research.[76] Here I have presented how feminist and poststructuralist challenges associated with big data, including guidance on establishing qualitative rigor, could be applied to big data. By illuminating the subject positions of those who control the black box gathering big data, I am hopeful that another step toward infiltrating the cyborg will be taken.[77]

Notes

1. Kitchin, "Big Data and Human Geography."
2. S. Smith, "Social Constructivisms."
3. boyd and Crawford, "Critical Questions for Big Data"; Dalton and Thatcher, "What Does a Critical Data Studies Look Like."
4. S. Smith, "Social Constructivisms."
5. Thatcher, "Big Data, Big Questions."
6. Elwood and Cope, "Introduction: Qualitative GIS"; Schuurman and Pratt, "Care of the Subject"; Schuurman, "Trouble in the Heartland"; Schuurman, GIS: A Short Introduction; Schuurman, "Database Ethnographies"; Schuurman, "Formalization Matters."
7. Baxter and Eyles, "Evaluating Qualitative Research"; England, "Getting Personal."
8. Rose, "Situating Knowledges."
9. Kitchin, "Big Data and Human Geography."
10. M. Graham and Shelton, "Geography and the Future of Big Data."
11. Barnes and Wilson, "Big Data, Social Physics"; Barnes, "Big Data, Little History."
12. Kitchin, "Big Data and Human Geography."
13. M. Graham and Shelton, "Geography and the Future of Big Data."
14. Kitchin, "Big Data and Human Geography."
15. Ricker, Hedley, and Daniel, "Fuzzy Boundaries."
16. Haklay, "Neogeography"; Hecht and Stephens, "Tale of Cities"; Kelley, "Urban Experience"; Stephens, "Gender and the GeoWeb"; Zook and Graham, "Mapping DigiPlace."

17. Crampton et al., "Beyond the Geotag"; M. Graham and Shelton, "Geography and the Future of Big Data"; Leszczynski and Wilson, "Guest Editorial"; Zook and Graham, "Mapping DigiPlace."
18. Wang and Armstrong, "Theoretical Approach."
19. See the work resulting from NSF Grant #1047916 SI2-SSI: CyberGIS Software Integration for Sustained Geospatial Innovation.
20. Buneman, Khanna, and Tan, "Data Provenance."
21. Thatcher, "Big Data, Big Questions."
22. Thatcher, "Big Data, Big Questions."
23. Thatcher, "Big Data, Big Questions."
24. Dalton and Thatcher, "What Does a Critical Data Studies Look Like"; Thatcher, "Big Data, Big Questions."
25. Baxter and Eyles, "Evaluating Qualitative Research."
26. Baxter and Eyles, "Evaluating Qualitative Research"; England, "Getting Personal."
27. England, "Getting Personal."
28. Baxter and Eyles, "Evaluating Qualitative Research"; Schuurman and Pratt, "Care of the Subject"; Wilson, "Towards a Genealogy."
29. England, "Getting Personal."
30. Baxter and Eyles, "Evaluating Qualitative Research."
31. Baxter and Eyles, "Evaluating Qualitative Research."
32. Pavlovskaya and St. Martin, "Feminism and Geographic Information Systems."
33. Leszczynski and Wilson, "Guest Editorial"; Schuurman and Pratt, "Care of the Subject"; Schuurman, "Trouble in the Heartland"; Schuurman, "Formalization Matters"; Wilson, "Towards a Genealogy."
34. M. Graham and Shelton, "Geography and the Future of Big Data."
35. Elwood and Cope, "Introduction: Qualitative GIS"; Schuurman and Pratt, "Care of the Subject"; Schuurman, "Trouble in the Heartland"; Schuurman, GIS: A Short Introduction; Wilson, "Towards a Genealogy."
36. Haraway, Cyborg Manifesto; Pickles, Ground Truth; Pavlovskaya and St. Martin, "Feminism and Geographic Information Systems"; Rose, "Situating Knowledges."
37. Pavlovskaya and St. Martin, "Feminism and Geographic Information Systems."
38. Elwood and Cope, "Introduction: Qualitative GIS"; Elwood and Leszczynski, "Privacy, Reconsidered"; Harvey, "To Volunteer or to Contribute"; Kwan, "Gender and Individual Access"; Leszczynski and Wilson, "Guest Editorial."
39. Haklay, "Neogeography"; Hecht and Stephens, "Tale of Cities"; Stephens, "Gender and the GeoWeb."
40. Schuurman and Pratt, "Care of the Subject"; Schuurman, "Formalization Matters."

41. Takacs, "How Does Your Positionality Bias Your Epistemology?"
42. Rose, "Situating Knowledges."
43. Elwood and Cope, "Introduction: Qualitative GIS."
44. Haraway, *Cyborg Manifesto*; Pavlovskaya and St. Martin, "Feminism and Geographic Information Systems."
45. Rocheleau, "Maps, Numbers, Text, and Context."
46. Harding, "After the Neutrality Ideal."
47. Rocheleau, "Maps, Numbers, Text, and Context."
48. Schuurman, *GIS: A Short Introduction*, 30.
49. Elwood and Cope, "Introduction: Qualitative GIS."
50. Crampton and Krygier, "Introduction to Critical Cartography"; Leszczynski and Elwood, "Feminist Geographies of New Spatial Media."
51. Takacs, "How Does Your Positionality Bias Your Epistemology?"
52. Elwood and Cope, "Introduction: Qualitative GIS"; Schuurman and Pratt, "Care of the Subject"; Schuurman, "Trouble in the Heartland"; Schuurman, *GIS: A Short Introduction*; Schuurman, "Database Ethnographies"; Schuurman, "Formalization Matters."
53. Rocheleau, "Maps, Numbers, Text, and Context"; Pavlovskaya and St. Martin, "Feminism and Geographic Information Systems."
54. Elwood, "Geographic Information Science" (parts I and II); Hecht and Stephens, "Tale of Cities."
55. Chambers, "Participatory Mapping and Geographic Information Systems."
56. M. Graham and Shelton, "Geography and the Future of Big Data."
57. Wilson, "Towards a Genealogy."
58. Wilson, "Towards a Genealogy."
59. Schuurman, *GIS: A Short Introduction*.
60. England, "Getting Personal."
61. M. Graham and Shelton, "Geography and the Future of Big Data"; Haraway, *Cyborg Manifesto*.
62. England, "Getting Personal."
63. Baxter and Eyles, "Evaluating Qualitative Research"; England, "Getting Personal."
64. Baxter and Eyles, "Evaluating Qualitative Research."
65. Baxter and Eyles, "Evaluating Qualitative Research."
66. M. Graham and Shelton, "Geography and the Future of Big Data"; Kitchin, "Big Data and Human Geography."
67. England, "Getting Personal."
68. England, "Getting Personal."
69. Rose, "Situating Knowledges"; Nast, "Women in the Field."

70. England, "Getting Personal."
71. England, "Getting Personal"; Rose, "Situating Knowledges."
72. England, "Getting Personal."
73. Dalton and Thatcher, "What Does a Critical Data Studies Look Like."
74. Baxter and Eyles, "Evaluating Qualitative Research"; England, "Getting Personal."
75. Kitchin, "Big Data and Human Geography"; Thatcher, "Big Data, Big Questions."
76. Elwood and Cope, "Introduction: Qualitative GIS"; Schuurman and Pratt, "Care of the Subject"; Schuurman, "Trouble in the Heartland"; Schuurman, GIS: A Short Introduction; Wilson, "Towards a Genealogy."
77. Haraway, Cyborg Manifesto.

PART 3
Empirical Interventions

5   A Hybrid Approach to Geotweets

Reading and Mapping Tweet Contexts on Marijuana Legalization and Same-Sex Marriage in Seattle, Washington

Jin-Kyu Jung and Jungyeop Shin

> Rigorous scientific scholarship is not about the one best method
> and universal truth. Rather, it is about an ongoing engagement
> and debate between the full range of "scientific" methods, each
> capable of producing reliable but situated understandings of the
> world, and involving the full range of potential stakeholders in
> seeking to understand and improve the world.
>
> —ERIC SHEPPARD, "Knowledge Production through Critical GIS"

We are living in a world of big data, in which 2.5 quintillion bytes of data are created every day, and more data are generated every two days than humankind created in all of history prior to 2003.[1] Some have estimated that in 2016 the size of generated data in the world will surpass not the petabyte or the exabyte but the zettabyte.[2] This would be equivalent to the amount of data that would result from every person in the United States taking a digital photo every second of every day for a month.[3] Big data is often characterized not only by its size or volume but also by its variety and velocity, such that traditional tools and processes cannot handle the data.[4] The responses to big data are varied, from exuberant excitement to extreme fear. For instance, the U.S. government and private corporations like IBM believe that big data can enhance accountability, boost economic productivity and government services, and even protect the rights of citizens.[5] Others fear that big data could erode privacy, diminish our critical perspective on data, diminish the use of finer-level small data, and eventually give us the fallacy that bigger data sets are always better data sets.[6]

These spectra of reactions to big data phenomena provide unique opportunities as well as challenges for geographers and GIScientists. While faster access to large and diverse data sets may offer unprecedented opportunities, it also presents methodological conundrums. From a data science perspective the proportion of data we can analyze is *naturally* declining (even though the amounts of data we can analyze are increasing), because the evolution of tools and processes cannot keep pace with the exponentially increasing numbers of data. However, the bigger issue facing researchers is the development of effective and appropriate methods capable of extracting meaningful information from big data. More specifically we are presented with the question of how we can analyze and interpret data in a way that preserves context and considers not only the size of a data set but also its depth. In researching big data we need to consider, borrowing Manovich's terms, both "deep data," often associated with rich descriptive qualitative data, and "surface data," associated with large volumes of quantitative data.[7] Even though many researchers have started to utilize big data, we have seen only a few examples of research that uses and takes into account both deep and surface data, or what Sui and Goodchild have characterized as the data avalanche: deep data for and about many.[8]

The main objective of our research is to develop a hybrid approach to explore, analyze, and represent big *spatial* data in a way that leverages both quantitative and qualitative methods throughout the research process. In particular this paper is a response to a general lack of approaches integrating qualitative ethnographic and interpretive methods with quantitative geographic visualization and spatio-temporal analyses. As Eric Sheppard suggests in our chapter epigraph, we would like to find a way to adopt and integrate different methodologies in meaningful ways to study big spatial data. This is also our attempt to revisit *and* advance the insights of innovative researchers, who have previously sought to integrate different methodological approaches in creative ways in order to develop mixed or hybrid approaches to the world of big spatial data.[9]

Jung and Shin

To provide an empirical example of our hybrid framework, we focus on spatially linked social media, in the form of geolocated tweets or geotweets. Geotweets are one important form of big data that have resulted from the growth of social, location, and mobile applications, or "SoLoMo."[10] In SoLoMo we see countless forms of user-generated content (e.g., photos), endless communication (e.g., tweets), user-defined key words or metadata (e.g., hashtags), and various location-based services (LBS). Location-based information is often one of the most important new features of mainstream social media, and it has given geographers a reason to pay greater attention to social media.[11] For instance, a search for the keyword "Twitter" on Google Scholar (scholar.google.com) results in more than 5.8 million papers as of December 10, 2015. Nevertheless, compared to other research areas related to tweets and social network analysis, relatively little research has focused explicitly on spatial or locational social data.[12]

In the following section we review existing geographical research involving geotweets. This review presents some key themes emerging from the various types of quantitative methods currently used to research tweets, and it will also reveal the relatively limited use of qualitative approaches. We further argue for the need to develop a hybrid approach that gives us a more holistic understanding of tweets. In the third section we briefly discuss the theory and practice of mixed-methods research, present our conceptual model of a hybrid approach, and situate our model within the discussion of mixed-methods research. Our model concurrently and sequentially integrates digital ethnography with spatio-temporal analyses and geographic visualization. The fourth section introduces our case study, which maps and examines three months of geotweet data in King County, Washington, related to the legalization of marijuana and same-sex marriage in Washington State in the 2012 general election. While this case study may not show a full picture of the hybrid approach, we hope that our focus on the process of building a hybrid approach will stimulate further discussions about applying such integrated methods to the geographic study of geotweets and big spatial data in general.

Existing Geographic Research on Geotweets

By using geotweets we can understand activities and opinions in virtual space, as well as their relationship to real world space and activities. With newly joined locational information added to the other obvious merits of Twitter data, geotweets are important sources of information for geographical research. We find that current research on geotweets has focused on three main themes: disaster, public health, and politics.

First, issues related to natural disasters, such as earthquakes, floods, and hurricanes, are critical to the safety of our lives. Compared to traditional, official data sets, tweets are more effective at identifying disasters and increasing awareness. Kent and Capello Jr. and Vieweg et al., for instance, have explored the correlation as well as the causal relationship between the location of a wildfire and the locations of fire-related tweets.[13] There have also been studies that have attempted to mine seismological data from sentiments and opinions expressed on Twitter.[14] The second theme is public health. Health issues, such as the occurrence and dispersion of epidemic diseases, have spatio-temporal perspectives. Official data sources are often slow to detect the propagation of diseases or the public's feelings about newly emerging diseases. For tracking and controlling disease and health issues, Twitter data have become more important.[15] The third theme is related to political issues and elections. There have been many research projects that have attempted to predict election results for political and social ballot measures using tweets. In many cases a significant correlation was identified between the contents of tweets and election results, thereby evaluating the prediction power of social media.[16] This has been implemented not only in the United States but also across the globe in locations such as Singapore and the Netherlands.[17] Many empirical studies have also proven that a strong positive correlation exists between Twitter data and election results.[18] Some researchers, however, dispute the correlation between Twitter data and elections results because of potential discrepancies between the two.[19] These discrepancies may be due to a lack of the accuracy necessary for prediction, overestimation of the election voting rate from rates of support on Twitter, and inconsistencies between opinions expressed in tweets and election support. This makes us wonder

Jung and Shin

if the tweets' contents could be used for predicting the political results or if we could find debates or activities in social media in relation to a political event like an election.

In methodological terms quantitative analyses are useful for the study of the distribution and patterns of tweets, as well as for finding the correlation between tweets and other variables. We consider three particular quantitative methods that can be used to explore the geographic context of tweets: geographic visualization, temporal visualization, and spatial analysis.[20]

The first methodology is geographic visualization. Many researchers have focused on finding out who is and who is not using social media, why and for what purposes they use it, and what visualization techniques greatly help such research exploration.[21] Geotweet data allow us to use the power of *geographic* visualization. Tsou et al. developed a kind of packed solution called VISION (visualizing information space in ontological network) to analyze the spatial distribution of webpages and content in social media, including data found via commercial search engines (e.g., Yahoo, Bing), the Twitter API, IP geolocation methods, and basic geographic information system (GIS) functions.[22] They created tweet density maps, tweet key word clouds, and spatio-temporal maps of geotweets for the 2012 U.S. presidential election. MacEachren et al. also suggested a total software solution for geotweets.[23] Their package, called SensePlace2, has several functions, such as collecting, processing, and visualizing tweets through mapping and interactive query. Both VISION and SensePlace2 are good examples of *geo*-visualizations, which capitalize on the geographic information associated with tweets.

The second methodology is temporal visualization. The temporal distribution of tweets and linked multimedia, such as photographs, is visualized using different temporal scales.[24] Researchers have created temporal visualizations at both local and global levels, in order to show the mobility patterns of geotweets at different times.[25] The temporal attributes of Twitter data can be explored using different temporal scales, such as monthly, daily, and hourly scales, and they are often visualized using histograms, frequency word clouds, and Twitter data clocks. Combined with geographic visualization, temporal visualization contributes to the creation of spatio-temporal visualizations of tweets.

The third type of methodology is based on traditional GIS spatial analysis and is used to find deeper geographic patterns within Twitter data. A series of spatial statistical analyses can be useful here. For example, for the individual point distribution of tweets, we can measure the density of point distributions, degree of clustering, and even the causal relationship between Twitter data and related socioeconomic data. Examples include the visualization of statistical deviational ellipse of point distributions, the extraction of information about obesity and health from tweets for comparison with the location of McDonald's restaurants, and the use of various regression and hot-spot analyses to explain tweet distributions in relation to various socioeconomic variables.[26] These examples all demonstrate innovative attempts to study geographic aspects of tweets through traditional GIS and spatial analyses.

Mixed-Method Research Design and Digital Ethnography

Diverse and creative efforts have been made to expand the geographical analysis of geotweets and social media in general. We have moved beyond just plotting dots (tweet points) on maps and have started to make good progress toward drawing meaningful quantitative information from tweets. However, existing discussions of geographic analysis of tweets have not fully considered the qualitative forms of Twitter data, and as a result researchers have not often adopted qualitative research methods in their analysis. We can create a hybrid or mixed-methods approach by mixing both quantitative and qualitative methods and more specifically by integrating digital ethnography with GIS and spatial analyses. Before we present the details of the proposed hybrid approach, we briefly discuss the mixed-methods approach—what it is and what its characteristics are.

Mixed-methods research supports the idea of combining quantitative and qualitative approaches in the methodology of research. The central promise of mixed-methods research is that different research methods offer possible solutions for one another's problems.[27] Different methods

Jung and Shin

not only cross-check each other but also improve and build upon one another. Taking a pragmatic approach, mixed-methods research recognizes that the research question is more important than either the method or the worldview that is supposed to underlie the method. In other words this approach no longer considers research methods to be extensions of epistemology and ontology, thereby preventing the research from being confined to any particular type of method.[28] In the discipline of geography both qualitative and quantitative researchers have begun to recognize the validity of mixed-methods approaches.[29] In particular the discussions surrounding mixed-methods research provided crucial ground for the development of qualitative GIS, which mainly looks at the possible methodological blending of GIS with qualitative data and research. Researchers have increasingly attempted to expand the potential of GIS for nonpositivist modes of analysis, including qualitative research, as part of a wider effort to problematize the contemporary divisions between epistemology, ontology, and methodology. They have made consistent efforts to break the taken-for-granted binaries of qualitative geography and quantitative geography and have problematized the narrowly defined meaning of GIS as a quantitative/positivistic research tool.[30]

The important discussions concern qualitative data. Qualitative data are considered to be a source of well-grounded, rich descriptions of processes and places, and they are usually in forms such as text, image, audio, and video rather than in numerical form.[31] Qualitative data are often collected from the field so that they reflect the everyday lives of people and society. Because qualitative data are traditionally close to a specific situation, it is believed that they better preserve contextual information related to the phenomenon under examination.[32] We suggest that geotweets can be considered qualitative data and that meaningful insights from tweets can be pulled out through *qualitative* research methods. However, there is a clear distinction between traditional qualitative data from the physical field and tweets. Geotweets are in digital or virtual space. For this reason it is important for us to think about a new type of qualitative research method for geotweets.

We critically examine the possibility of using what Murthy calls "digital ethnography."[33] Digital ethnography is the *ethnographic* study of or with *digital* objects and technologies, in which we systematically study people, places, and cultures using digital artifacts. Murthy argues that research done exclusively online, for instance, on social networking websites, blogs, and digital video, "can be highly fruitful," even though we cannot consider digital ethnography to be superior to traditional ethnographic research.[34] We agree with him that digital data are good sources for hearing and telling nuanced, situated, and contextualized stories if we approach and treat them with the same rigor, integrity, and ethics that we commonly use in traditional ethnographic research. Ethnography with digital data has the same value as nondigital ethnography. The increased volumes of not only textual data but also multimedia such as audio, video, and photos make digital ethnography an attractive qualitative method. We are already starting to see the practical implications of digital ethnography, as evidenced by the use of visual methodologies by planners and researchers.[35] Digital ethnography can be an *interpretive* and an *analytical* qualitative method that is capable of making sense of phenomena in tweets. Digital ethnography can also help us find out the meanings people (e.g., tweeters) bring to phenomena, which are the stories, lives, opinions, and expressions "posted" in their tweets. Now the question is how we might mix digital ethnography with quantitative methods in a hybrid research framework.

Methods can be mixed in different ways, and mixed-methods research design can therefore be classified by the priority given to either quantitative or qualitative methods, by the procedures adopted to mix different methods, and by the level at which methods are integrated (see table 5-1).[36] Hybrid methods can prioritize qualitative (QUAL) or quantitative (QUAN) methods, or they can give an equal weight to each type of method (QUAL + QUAN). Both qualitative and quantitative methods can be used in parallel (simultaneous), or one method can be used first in order to develop measures and outcomes for the following method (sequential). In addition the hybridization can occur at the level of data, method, and even theory, even though theory-level integration might be the most challenging task.

Jung and Shin

TABLE 5-1. Types and levels of hybridity

	HYBRIDITY
Priority	QUAL, QUAN, or QUAL + QUAN
Procedure	Simultaneous or sequential
Integration	Data level, method level, or theory level

Figure 5-1 shows the conceptual model and process of the hybrid approach upon which we are building. It demonstrates what, how, and when quantitative and qualitative methods and analyses are used and integrated in our research process. We will explain each step in detail with our case study in the next section.

Case Study: Building a Hybrid Approach for Analyzing Geotweets

Step 1: Data Collection

The study area for our project was King County, which is the most populous county in Washington State and the county in which Seattle is located. Geotweet data were collected from King County over the three-month period from August 1 to October 31, 2012.[37] A total of 818,818 tweets were collected, and the resulting data set contains detailed information, including user IDs, user descriptions, geographic coordinates, date and time stamps, place types, and the main text of tweets. We used the geographic coordinates of each tweet to generate tweet maps and then used Esri ArcGIS Desktop Suite for spatio-temporal analyses and geographic visualization. All data were stored in table format (Excel) for searching, retrieving, and analyzing tweet information. We also used computer-assisted qualitative data analysis software (CAQDAS), especially ATLAS.Ti 6.0, to facilitate qualitative data analysis. Geotweets include both quantitative and qualitative data, which presents data-level hybridity.

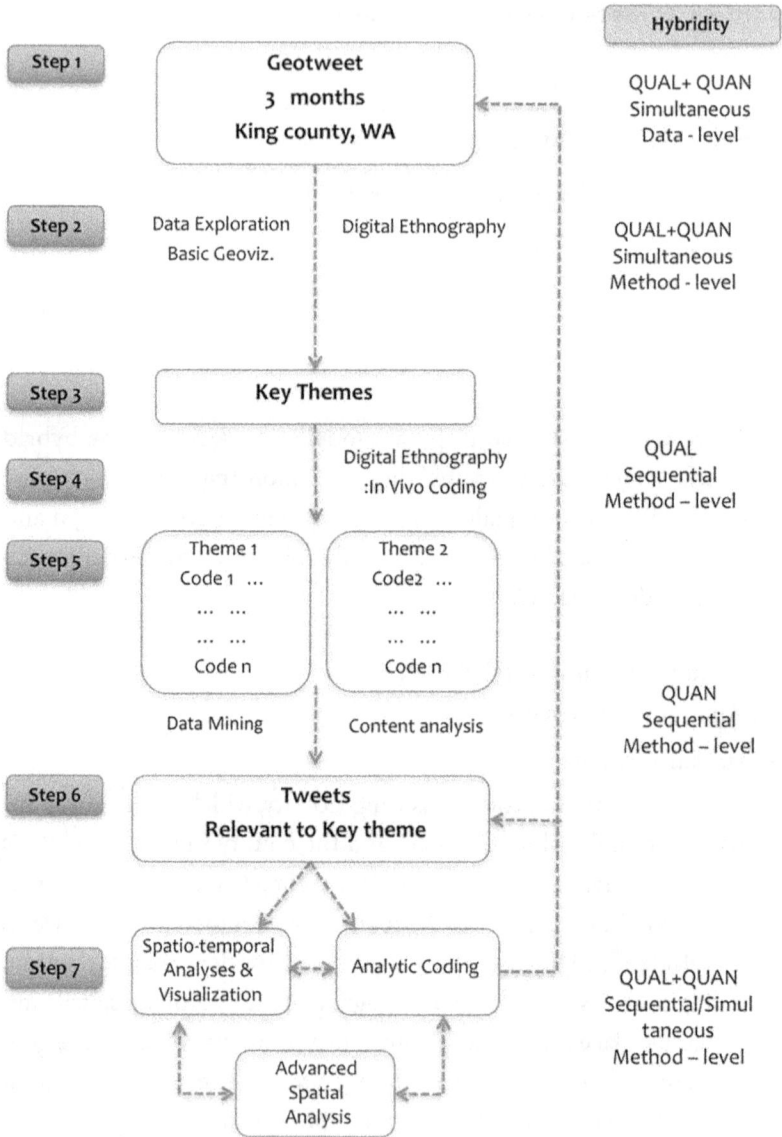

FIG. 5-1. Conceptual model of hybrid approach. Created by J-K. Jung and J. Shin.

Step 2: Data Exploration and Basic Visualization

We began by visualizing the basic spatial and temporal distributions of all of the tweets. Mapping close to one million tweets within King County did not show any identifiable or meaningful spatial patterns and characteristics except that most tweets were from the western part of the county, especially around Seattle. We also conducted a simple content analysis by counting the frequency of words, and we visualized the tweets' contents by creating content clouds. Although these were not made for advanced textual analysis or visualization, they were useful for data exploration and initial visualization that offers a quick overview of the data set. In parallel we began digital ethnographic research by reading and exploring multiple possible meanings of the original tweets' texts. Certain themes naturally emerged in the process. We identified the most tweets about the following issues/topics: the Summer Olympics, the U.S. swimming team (especially Michael Phelps), the NFL, the Seahawks, foods, restaurants, crimes, bars, music, singers, weather, friends, and elections. In particular the tweets about political issues and elections caught our attention, because there were many tweets about the upcoming presidential election contests between Barack Obama and Mitt Romney, as well as about the 2012 election at the state level. In this step of our analysis both quantitative and qualitative methods were simultaneously used in the process of exploring and finding key themes for the research, and the integration occurred at the method level.

Step 3: Selecting Key Themes

Because of the sensitivity of the issues and the intensifying interests and controversy we could identify in the public from our initial readings of tweet texts in the previous step, we decided to focus on two particular local agendas in the Washington State election: Initiative 502 (I-502), related to marijuana legalization, and Referendum 74 (R-74), related to same-sex marriage. We also set our overarching research goal to be the identification of any meaningful relationship that might exist between geotweets and the 2012 state-level election, especially I-502 and R-74.

To answer this question we decided to look at any potential correlation between the geotweet and socioeconomic or demographic variables, as well as between the geotweets and public opinion as represented by local mainstream media coverage (e.g., *Seattle Times*). To measure any potential correlation we also acquired the election results data, including a precinct-level GIS file. The process of choosing key themes from the earlier two steps in the study helped us to generate key research questions and also made us consider other variables with which to work.

Step 4: Digital Ethnography: In Vivo Coding

The digital ethnographic research we conducted with geotweets was based on grounded theory, especially the coding methods used.[38] Grounded theory is a method used to inductively generate theories from empirical data that have been systematically gathered. The coding process is an integral part of grounded theory. Coding is a way of understanding the contextual meaning of qualitative data, and codes are categories or keywords that represent the context of qualitative data.[39] There are generally two broad levels in the coding process, both of which we implemented in our research. One is in vivo coding, which can be thought of as being emic to the original data. This is based on the idea that codes should typically not be pre-defined by researchers but instead should emerge from the data.[40] For this reason in vivo codes are often words or phrases taken directly from the texts. On the other hand analytic coding is another level of coding process, one in which researchers generate more interpretive and reflective codes.

During step 4 of the research process we primarily performed in vivo coding, paying particular attention to our two key themes (i.e., I-502 and R-74). By carefully reading the main content of tweets one by one, we generated codes directly from the tweets' texts. For instance, in relation to I-502 if "marijuana" was included in a tweet, the word "marijuana" became one of our in vivo codes. In the initial coding process in vivo codes were not produced as a final outcome of the research but were instead intended to be used to search for and find tweets that might be related to the two themes we focused on in the following analysis. In other words the initial

Jung and Shin

qualitative research, in this case in vivo coding, was used for identifying tweets that may be more contextually related to our two focused themes. Ideally it would have been best if we could have read all tweets and completed the in vivo coding for all the tweets we had. However, this was impossible and also unrealistic because of the number of tweets and the time it took to code each tweet.[41] We therefore read and coded a total of 24,560 tweets one by one, which was a random selection of 3 percent of all gathered tweets.

In terms of hybridity this step was mainly based on qualitative research at the method level, and it was a sequential integration because its results were designed to help the procedure of the following quantitative research.

Step 5: Codes Related to Key Themes

We generated codes related to I-502 and R-74 by considering both the tweet text and the hashtag fields in the database. We purposefully created four "selected codes" for indicating the association with and the nature of support for the two themes, I-502 and R-74. We added the following four new fields (columns) in the database: "I-502" (tweets related to marijuana legalization), "R-74" (tweets related to R-74, marriage equality, or same-sex marriage), "I-502 support" (tweets showing support or no support for I-502), and "R-74 support" (tweets showing support or no support for R-74).

The following codes were created in relation to I-502: "marijuana/ marihuana," "BHO (Butane Hash Oil)," "weed," "haze," "pot," "hempfest," "legalize," "catnip," "medical," "502 yes," "pop," "dope," "hufplantlife," "doit," "potpeeve," "vote," and "502."[42] Codes associated with R-74 were the following: "Referendum 74," "same sex marriage," "gay marriage," "marriage equality," "marriage right," "approve 74," "WA4M," "R74," "ApproveR74," "vote4love," "Seattle approve R74," "gay sorority," "R74 marriage," "loveislove," "freedomtomarry," and "Approve 74."

The initial coding process also helped us see various feelings, sentiments, and opinions about the two election agendas as evidenced by the following tweets:

"People trip about weed like me smoking it is gonna dramatically effect their lives"

"It's an interesting thought, but the poor disappearing [in the conversation of] has sinister implications."

"See it's hot in Seattle people. We got the 3ws too, women weed and weather"

"King4News poll: widespread support for legal marijuana, same-sex marriage:http://t.co/nyL1*** check that out"

"@cityclublive@UWComm Marijuana Legalization 1-502 Same Sex Marriage R-74 Debate http://t.co/tsNJ!***"[43]

Surprisingly, there were also many (un)expected examples describing criminal activities (e.g., a dealer kept updating his/her current location so that people could find and buy marijuana), slang or code words referring to marijuana (e.g., green, pop, BHO [butane hash oil] made from marijuana), and the organization of illegal activity. For instance, one of the most prolific participants in the Twitter discussions of marijuana legalization turned out to be one of the organizers/vendors of the Seattle Hempfest 2012, where she could sell what she claimed to be organic (marijuana) products. Her linked Flickr and Facebook pages showcased a full set of photos of the products she carries. Also it was not difficult to identify the Twitter IDs or hashtags of popular marijuana distributors, some examples being mary_jane and kusharmy for the former and #canabiscup, #CANNACig, #legalizemeth, #highttimes, and #addition among the latter.[44]

Step 6: Data Mining

With all the generated codes we extracted tweets that were particularly related to each key theme. Then quantitative content analysis was used to search for tweets relevant to the key themes. We searched for tweets containing particular words (e.g., the in vivo codes). Content analysis also allowed us to quantitatively measure the frequency of words. In this step only a quantitative method was used in a sequential order at the method level.

TABLE 5-2. Total number of supportive/nonsupportive tweets for I-502 and R-74

	I-502 (MARIJUANA LEGALIZATION)	R-74 (SAME-SEX MARRIAGE)
Total number of tweets	1,071	400
Support/nonsupport	Yes (35%), no (18.5%), neutral (46.5%)	Yes (64.2%), no (3.5%), neutral (32.3%)
Election results	Yes (61.7%), no (33.5%), null/invalid votes (2.8%)	Yes (65.3%), no (32.1%), neutral (2.5%)

Table 5-2 summarizes the total number of tweets, the number of supportive/nonsupportive tweets for I-502 and R-74, and the election results for both agendas. The number of tweets related to I-502 and R-74 was quite small compared to the total number of tweets. Small numbers of related tweets might also prove the lack of local residents' interest in the political issues, even though the issues seemed to be quite controversial and a popular topic of discussion in the main media coverage. The number of tweets for I-502 was almost double the figure for R-74. However, we also found that we cannot clearly determine whether many of the tweets for I-502 and R-74 were supportive or nonsupportive of each agenda; we categorized those as "neutral." Neutral tweets were more common for I-502, and many tweets that we extracted from earlier steps were not explicitly about I-502. While the majority of tweets were in support of R-74, only 35 percent of tweets showed support for I-502. This also suggested that some topics with wider cultural variation (e.g., cannabis use) may be more easily mistaken than explicit political expression (e.g., support for legalization).

The total number of registered voters was 1,165,084, and about 84 percent of them voted; the total number of votes cast was 973,894. Each issue had the same number of votes because the votes occurred during the same election on the same day. The rate of support for I-502 in the election (61.7

percent) was almost double the rate of support in tweets (35 percent) for the initiative. However, the rate of support for R-74 (65.3 percent) in the election was similar to the support in tweets (64.2 percent) for the reform. The comparison between a real election result and the key themes and codes discussed in the tweets allowed us to make connections between digital (virtual) and place-based physical realities.

Step 7: Basic Spatio-Temporal Analysis and Visualization

We applied various spatial and temporal analyses and also started to look at their correlation with other socioeconomic and demographic characteristics of King County, such as median household income and the distribution of youth. In these analyses we used only the tweets we extracted during the previous analysis.

Figures 5-3 through 5-6 show the geographic distributions of tweets for I-502 and R-74, the distribution of young people in their twenties and thirties, and the distribution of household median income. The spatial unit of these maps is the census tract. Figures 5-3 and 5-4 are density maps. To explore detailed spatial patterns, we overlaid the point locations of tweets with a kernel density surface. The kernel density was created using a bivariate probability density function, as given in figure 5-2.[45]

Kernel function K could have different forms, and in this analysis the quadratic function was used with 8,000 feet as a search radius and 1,000 feet as an output cell size.[46] The tweets for I-502 and R-74 are mostly distributed in the western part of King County, where the urban regions are located. The locations for R-74 are distributed much like the ones for I-502, and they are particularly concentrated in downtown Seattle. Even though the total number of tweets for R-74 is smaller than the figure for I-502, overall patterns are quite similar for both.

We also looked at socio-demographic information such as the distribution of young people in their twenties and thirties, as well as the median household income.[47] Young people are more clustered in urban regions centered on downtown Seattle, the University of Washington's Seattle campus, and Bellevue; their geographic distribution has a pattern quite

$$f(x) = \frac{1}{nh} \sum_{i=1}^{n} K\left(\frac{x - xi}{h}\right)$$

n = total number of points, h = the bandwidth to determine the amount of smoothing
K = kernel function, x = location of estimated point, x_i = location of known point i

FIG. 5-2. Bivariate probability density function to determine kernel density. Created by J-K. Jung and J. Shin.

similar to the distribution of tweets for I-502 and R-74. From an economic perspective the distribution of median household income is uneven (fig. 5-6). Incomes are higher in urban areas and much lower in rural areas, especially the eastern and southern parts of King County. The distributions of the tweets are similar to young people's distribution in the region but are different from the pattern of median income.

There are some temporal variations as well (fig. 5-7). The percentage of tweets on Sunday is slightly lower than on other days. Monday has the most tweet activity for I-502, while the midweek days are more active for R-74 tweets. If we look at the daily pattern of tweet activity in general, the least active period is from dawn to the morning hours (from about 5:00 to 10:00 a.m.). Tweet activity becomes more vigorous during the afternoon hours (from about 2:00 to 5:00 p.m.), and it reaches a peak during the evening and nighttime hours.

As we conducted basic spatio-temporal analyses and created maps, we simultaneously continued qualitative analysis, particularly analytic coding on the data set. As you may recall from our previous discussion of coding, analytic coding is a more interpretive process. It helps researchers understand the contextual meanings of qualitative data (e.g., tweets) that were often not as visible in the tweet texts or in vivo codes. Deeper reading and interpretation of only extracted tweets greatly helped us to identify and reveal hidden contexts. For instance, "equality," "family," and "kids and R-74" are examples of newly created analytic codes. There were

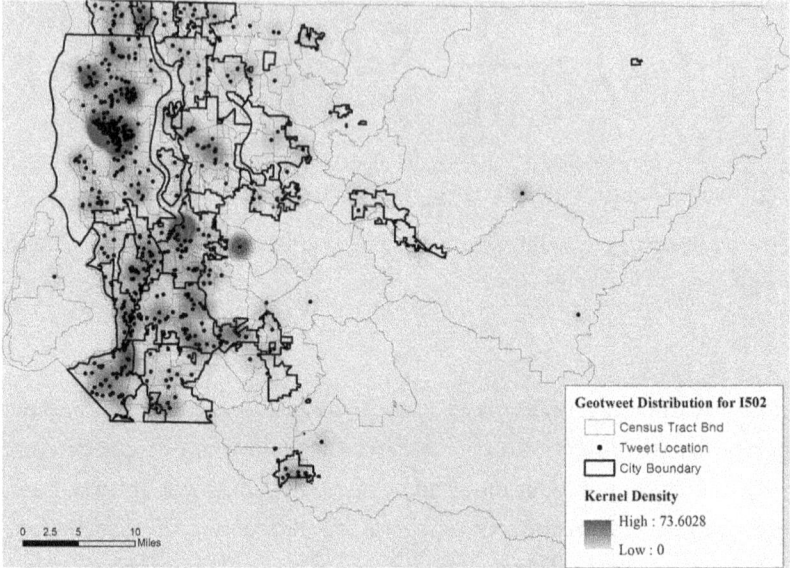

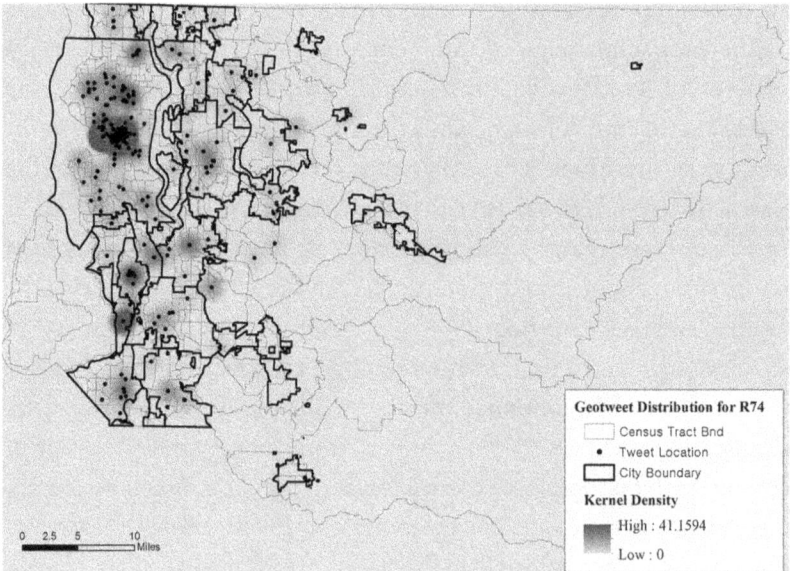

FIG. 5-3. (*top*) Spatial distribution of extracted tweets and distribution of young people and median income, tweet distribution for I-502. Created by J-K. Jung and J. Shin.

FIG. 5-4. (*bottom*) Spatial distribution of extracted tweets and distribution of young people and median income, tweet distribution for R-74. Created by J-K. Jung and J. Shin.

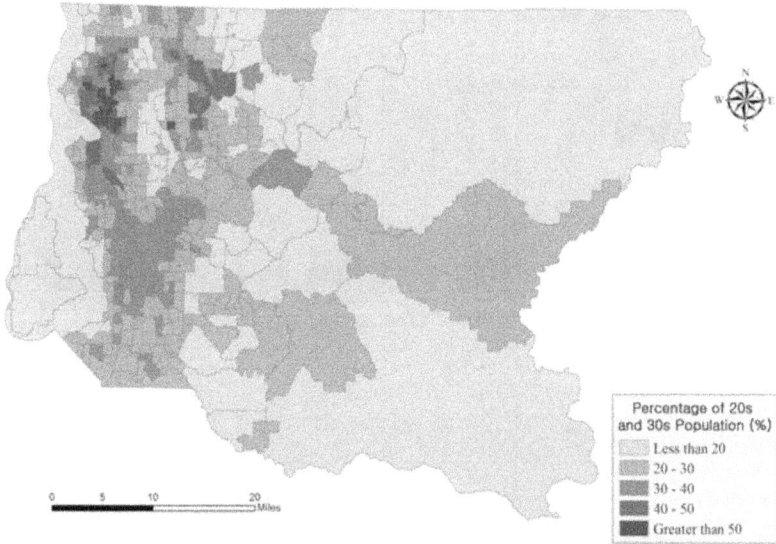

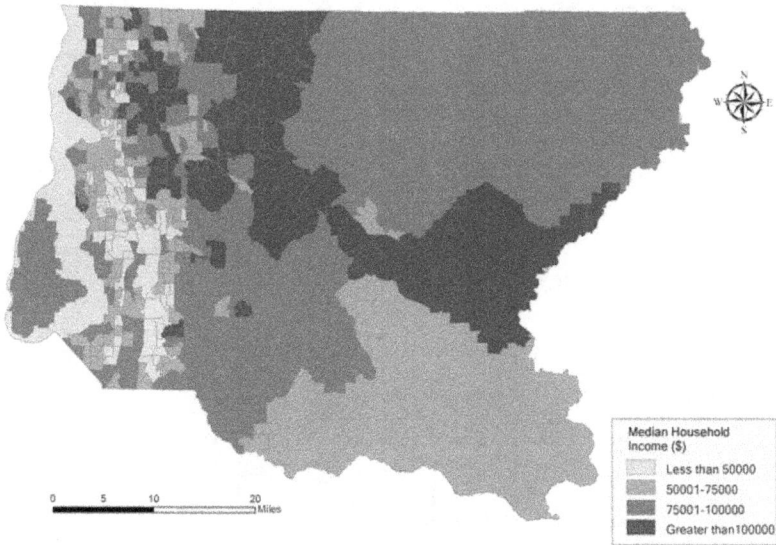

FIG. 5-5. (*top*) Spatial distribution of extracted tweets and distribution of young people and median income, distribution of young people (twenties and thirties). Created by J-K. Jung and J. Shin.

FIG. 5-6. (*bottom*) Spatial distribution of extracted tweets and distribution of young people and median income, distribution of household median income. Created by J-K. Jung and J. Shin.

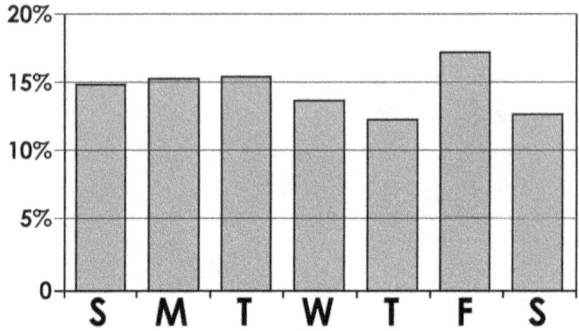

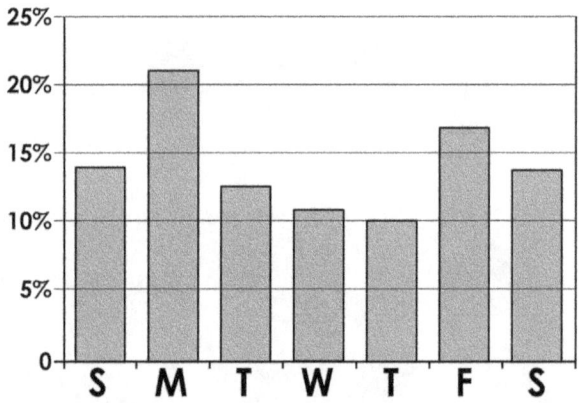

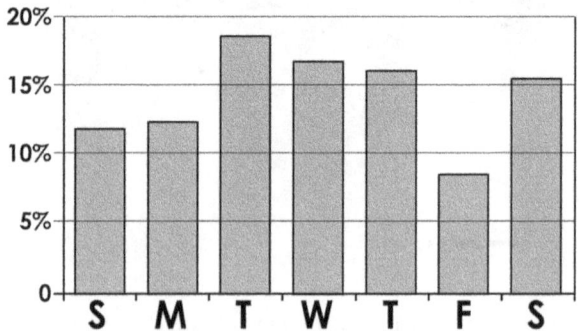

FIG. 5-7. Temporal distribution of tweets, weekly patterns. Created by J-K. Jung and J. Shin.

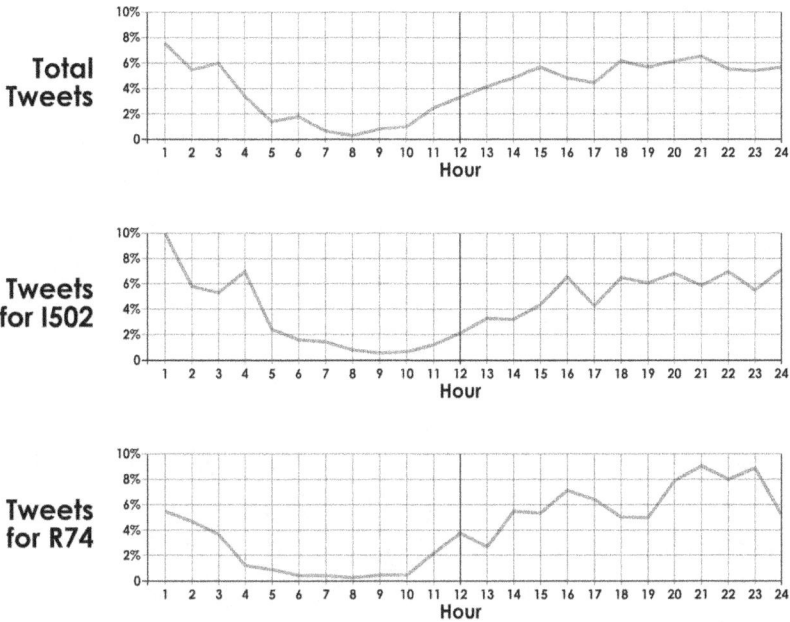

FIG. 5-8. Temporal distribution of tweets, daily patterns. Created by J-K. Jung and J. Shin.

debates regarding children's participation in the pro-I-502 rallies as well as R-74 protests. We are currently working through the analytic coding stage.

In order to explore the clustering of tweets we also implemented a hot-spot analysis using ArcGIS 10.1. Hot-spot analysis is a method for detecting hot-spot regions (those with a higher frequency of tweets) and "cold-spot" regions (with a lower frequency of tweets) based on spatial autocorrelation. For hot-spot analysis Getis-Ord Gi* was used for mapping clusters of tweets, as presented in figure 5-9.

Figures 5-10 through 5-12 shows the typical patterns of clusters for the tweets. The clusters for total tweets are shown in figure 5-10, which indicates that tweets are located mostly in the western part of King County. The hot-spot patterns of tweets for I-502 and R-74 are fairly similar to the pattern of the total tweets (figs. 5-11 and 5-12). However, there are also differences in terms of the locations of the hot spots between figures 5-11

$$G_i^* = \frac{\sum_{j=1}^{n} \omega_{i,j}\, x_j - \bar{X} \sum_{j=1}^{n} \omega_{i,j}}{S \sqrt{\dfrac{n \sum_{j=1} \omega_{ij}^2\, x_j - (\sum_{j=1} \omega_{ij})^2}{n-1}}}$$

x_i, x_j: attribute of region i, j, s: standard deviation, w_{ij}: spatial weight, n: number of region.

FIG. 5-9. Getis-Ord Gi* cluster determination. Created by J-K. Jung and J. Shin.

and 5-12. The hot spots for I-502 are shown with two black clusters below the city of Seattle, while the hot spot for R-74 is shown with one big black cluster around the city of Seattle. In comparison to the distribution of median income and young people (see fig. 5-6), the locations of hot spots seem to be positively correlated to the clusters of young people. On the other hand we observed that the concentration of high median income is more closely related to the locations of cold spots.

A correlation analysis was conducted to identify any positive or negative linear dependence between total tweets, socioeconomic variables, and voting results. The total number of tweets has a positive, relatively strong correlation with the percentage of young people in a population. In particular, populations of young people and tweets for I-502 show stronger positive correlation than populations of young people and tweets for R-74. This may be due to the fact that young people tend to be more interested in I-502 than R-74. On the other hand the correlation between young populations and median household income was moderate or even negative. It may explain a relatively weak interest in the two election agendas from higher-income groups. In terms of the support rates for both I-502 and R-74 no significant correlations were found with the two socioeconomic variables we explored.

More advanced analysis, including cluster analysis and spatial regression, can be used to explore the relationship between tweets and election

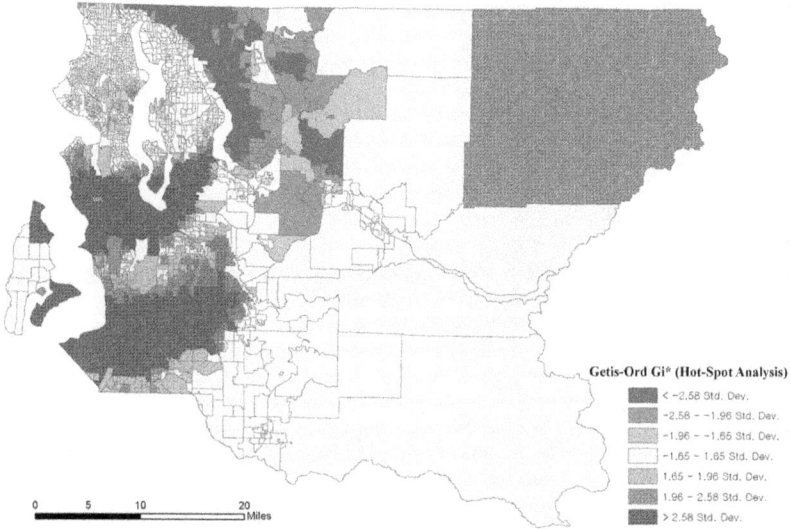

Getis-Ord Gi* (Hot-Spot Analysis)

■ < −2.58 Std. Dev.
■ −2.58 – −1.96 Std. Dev.
■ −1.96 – −1.65 Std. Dev.
□ −1.65 – 1.65 Std. Dev.
▨ 1.65 – 1.96 Std. Dev.
▨ 1.96 – 2.58 Std. Dev.
■ > 2.58 Std. Dev.

0   5   10   20 Miles

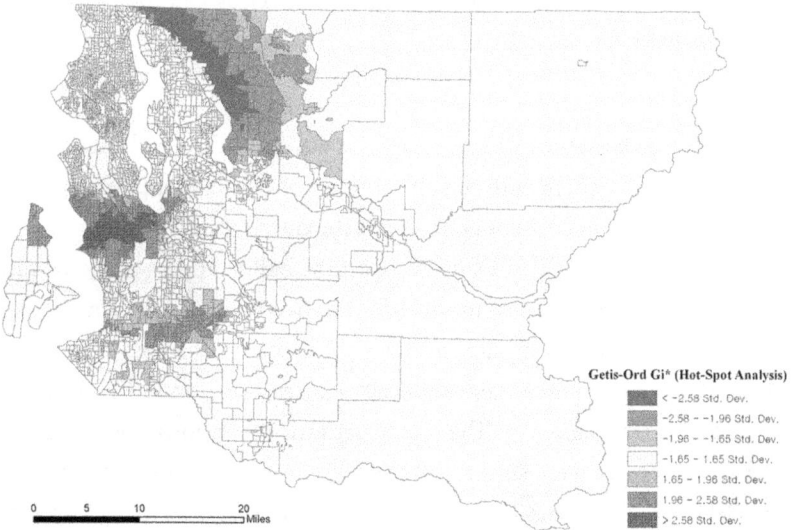

Getis-Ord Gi* (Hot-Spot Analysis)

■ < −2.58 Std. Dev.
■ −2.58 – −1.96 Std. Dev.
■ −1.96 – −1.65 Std. Dev.
□ −1.65 – 1.65 Std. Dev.
▨ 1.65 – 1.96 Std. Dev.
▨ 1.96 – 2.58 Std. Dev.
■ > 2.58 Std. Dev.

0   5   10   20 Miles

FIG. 5-10. (*top*) Hot-spot analysis of total tweets (Getis-Ord Gi*). Created by J-K. Jung and J. Shin.

FIG. 5-11. (*bottom*) Hot-spot analysis of tweets for I-502 issue (Getis-Ord Gi*). Created by J-K. Jung and J. Shin.

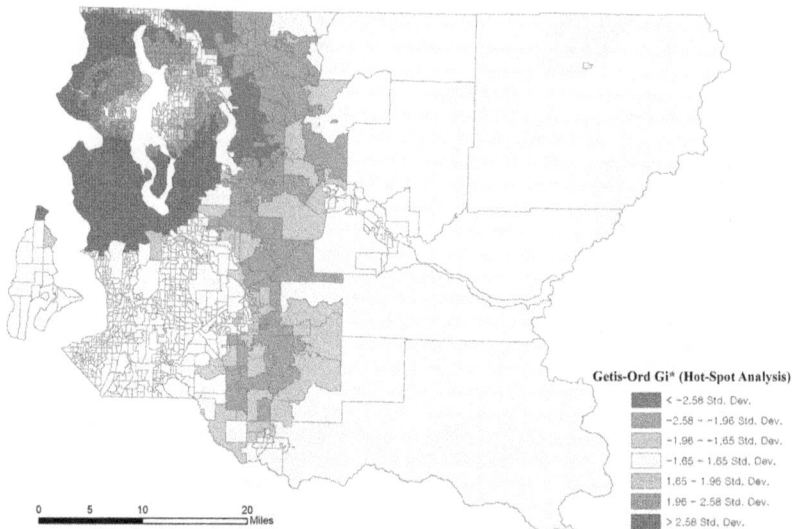

FIG. 5-12. Hot-spot analysis of tweets for R-74 issue (Getis-Ord Gi*). Created by J-K. Jung and J. Shin.

results. The precinct-level election results were gathered and analyzed in order to compare them with the distribution of tweets in support of I-502 and R-74. For instance, figure 5-13 shows the distribution of voter turnout in districts where I-502 and R-74 were on the ballot, and figures 5-14 and 5-15 show the percentage of support for the two ballot items. The overall percentage of votes cast in support for I-502 and R-74 in King County was 83.6, and the same percentage applies because voting on both measures took place on the same day. The spatial patterns of supportive voting rates for the two issues were similar (see figs. 5-14 and 5-15) and show higher rates of support around the city of Seattle and relatively lower support in the eastern part of the county. Interestingly the regions with support rates below 50 percent are larger for R-74 than for I-502.

We will continue seeking correlations or comprehensive information between tweets and election results as a way of juxtaposing qualitative and quantitative data and research methods. We will use GIS and spatial analyses in particular during the course of building hybrid approaches.

Jung and Shin

Voting Percentage
(Referendum 74)

■ < -2.5 Std. Dev.
■ -2.5 - -1.5 Std. Dev.
■ -1.5 - -0.50 Std. Dev.
☐ -0.50 - 0.50 Std. Dev.
■ 0.50 - 1.5 Std. Dev.
■ 1.5 - 2.1 Std. Dev.

0 5 10 20 Miles

Percent with "Yes"
in Initiative 502
☐ No Data
☐ Less than 50%
▨ 50 - 60 %
■ 60 - 70 %
■ 70 - 80 %
■ Greater than 80 %

0 5 10 20 Miles

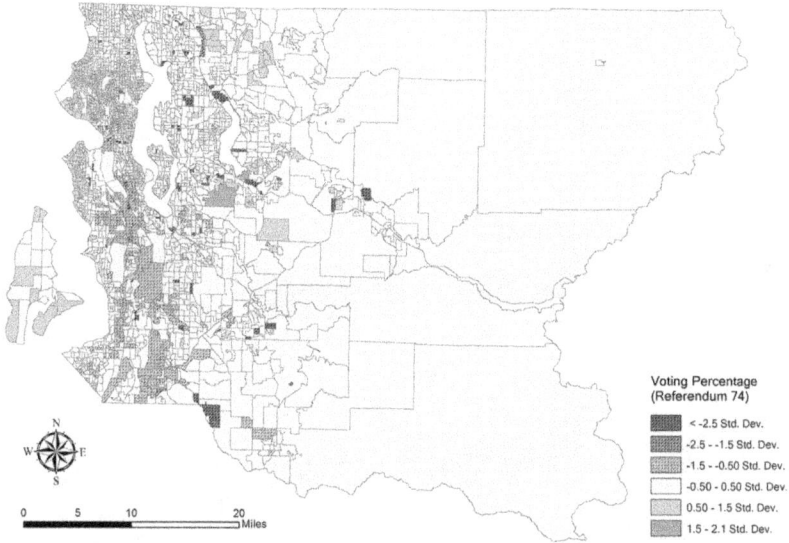

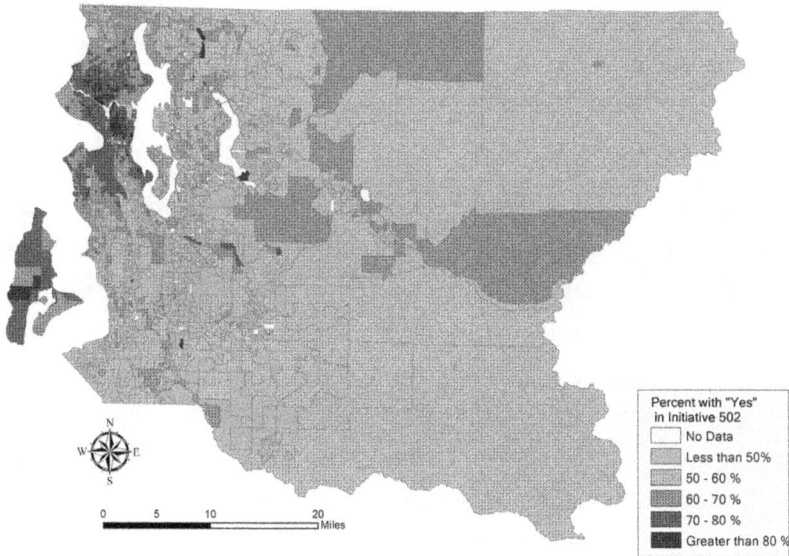

FIG. 5-13. (*top*) Spatial distribution of the voting rate for I-502 and R-74. Created by J-K. Jung and J. Shin.

FIG. 5-14. (*bottom*) Spatial distribution of the voting percentage favoring I-502. Created by J-K. Jung and J. Shin.

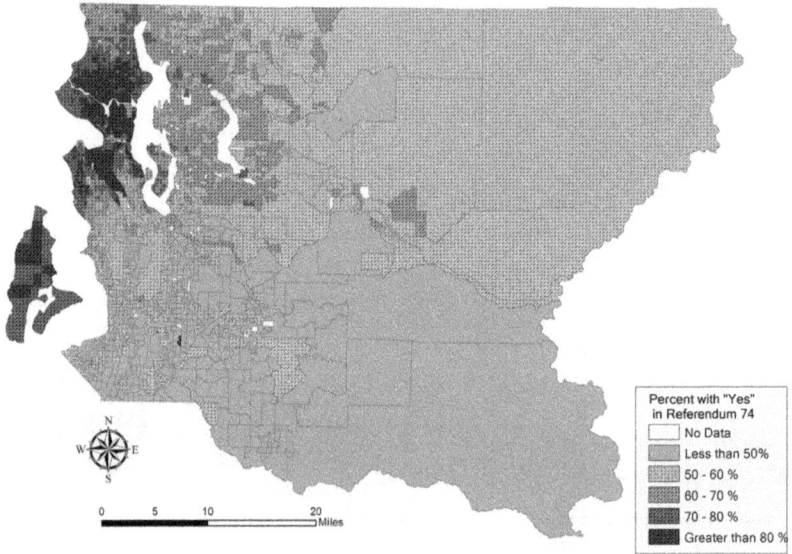

FIG. 5-15. Spatial distribution of the voting percentage favoring R-74. Created by J-K. Jung and J. Shin.

Conclusions

We have attempted to develop and apply a hybrid approach to study geotweets in a way that concurrently and sequentially integrates both quantitative and qualitative methods in the research process, in particular by combining digital ethnography with spatio-temporal analyses and visualization. We generated codes from initial digital ethnographic research and then searched for tweets that are relevant to the key topics of our investigation. We also applied various types of spatio-temporal analysis and geographic visualization, simultaneously as well as sequentially. This research was not intended to be prescriptive but rather a modest attempt to further the conversation on the development of effective hybrid methods capable of extracting "meanings" from big data. We began our research by asking what kinds of actions are taken to perform big data analyses, as well as by reviewing what kinds of geographical analyses have been used to study big spatial data. Many researchers, including geographers, seem

Jung and Shin

to have jumped on the bandwagon of big data with a naïve belief that it might lead them to promising outcomes without realizing the challenges of understanding and analyzing them. They might not even consider making any necessary connection to existing theory and disciplinary knowledge but only have enthusiasm about the prevalence of big data leading to completely new scientific findings.[48] What we worry about is not so much the separation of quantitative and qualitative research in the study of big data but the careless reduction that many big data analysts might impose on "critical" quantitative *and* qualitative research. After going through the various steps of developing a hybrid method, we have also experienced how difficult it is to take both the volume and depth of big data into consideration and how practically challenging it is to integrate both quantitative and qualitative data and research methods into big data analyses.

There are remaining methodological issues in developing a hybrid method. Our research could be regarded as incomplete from the perspective of either quantitative or qualitative analysis. Tweets, particularly geotweets, have inherent technical limitations to their use as a source for geographic research. Only 1 percent of all users account for more than 20 percent of all tweets, and no more than 3 percent of tweets are geo-referenced tweets.[49] However, reflections on the strengths and weaknesses of the quantitative spatial methods and qualitative ethnographic methods that we implemented in our geotweet analysis offer the new possibility of developing hybrid approaches that minimize the weaknesses and increase the strengths of both types of methods.

For instance, from the point of view of spatial analysts, it is still too early to conclude whether or not there is any positive or negative correlation between the location of tweets and what people tweet or, more closely related to our project, whether there is a correlation between what people tweet about the legalization of marijuana and same-sex marriage and the results of the election. Whether Twitter location really matters or not is a bigger question to answer. However, we also need to aim toward extracting meaningful geographical characteristics from tweet texts. For example, the results of correlation analysis with Pearson methods between three types of the tweets (supportive/nonsupportive/neutral) and socioeconomic factors, the election result, and spatial regression analysis can be considered together in order to explore the effects of socioeconomic factors on tweets. And while we have not included this in our research to date, we increasingly see a great potential for automatic, computational, and quantitative analysis using natural language processing and corpus linguistics techniques adopted from the field of machine learning.[50] Digital ethnography was useful to inform the 3 percent sample of the data set collected, and we generated codes and themes. From the terms used to generate those themes we may apply a supervised machine learning classification to classify the remaining 97 percent and identify words relevant to the themes. While we are aware of the fact that the common machine-learning algorithm may not fully consider the qualifying information in tweets, it offers a more automated and implementable analysis of tweet texts.

Following the perspective of digital ethnographers, we also valued each tweet as an important ethnographic piece of data and treated it as inherently meaningful. However, the fact that we could manually read and analyze no more than 250 tweets per hour was quite discouraging, considering our limited time and resources in comparison to the vast numbers of tweets we needed to review. We also realize a distinctive difference between traditional ethnographic and qualitative research and digital ethnography using tweets. In traditional ethnography the researchers are often embedded in the field, where they become active participants in everyday activities. Although we approached geotweets with the same rigor, integrity, and ethics that we use in any traditional ethnographic study, it was challenging to contextualize data and to make a connection to the realities. For example, it was difficult to tease out differences between "talk about marijuana" and "marijuana legalization" only from the tweets. This research offers a similar lesson to that articulated by Postill and Pink, who argue that big data ethnography should involve living part of one's life in the digital social media world and another part in the material, everyday world that involves participation, engagement, and collaboration in social media data and visualization.[51] This direct involvement should be based on the realization that social media has substantially altered how people render, discover, and "visualize" information in the social mediascape and that there is a need for an ethno(carto)graphy of the production of geographic visualization and GIS.[52]

In spite of these challenges we would like to close this chapter by reemphasizing the importance of intertwining qualitative methods with spatio-temporal analyses and visualization for the geographic analysis of tweets. It is important to know what people are saying and how topics are trending in the Twitter world. One powerful indicator showing the potential impact of Twitter's sentiment is the Twitter tweets per second (Ttps) index.[53] We can also measure the impact of tweets by the number of Twitter users following the tweeter. These are helpful quantitative indices, but they cannot effectively represent each individual tweeter's feelings or sentiments. There is the danger of overlooking small things by getting the bigger data.[54] On the other hand there is another type of danger in fetishizing numbers that cannot be fully considered or included in analysis.[55] We need an appropriate hybrid method for valuing and using both quantitative data (e.g., Ttps index) and qualitative data (e.g., tweet texts) in a contextually appropriate manner.

With hybrid methods each part of the research process contributes greatly to drawing more meaningful insights. Particularly in the era of big data we build hybrid- or mixed-methods approaches not only to answer the main research question in various and creative ways but also to generate and discover relevant new questions that motivate us. This is very different from pragmatism, which is considered to be an attractive philosophical partner of a mixed-methods approach.[56] We can go beyond the idea that research approaches should be mixed in ways that offer the best opportunities for answering the main research question. We can gain new insights from the process if we continue to put more effort into building a more comprehensive geographic approach to studying geotweets and big data in general. This gives us another reason to take the context of tweets into account and to build hybrid research frameworks capable of providing more contextualized pictures of the "small deep data" in big data. We hope our project suggests one form of "hybrid geographies" to foster a methodological mixing of quantitative and qualitative methods.[57]

A hybrid framework provides a new way to reflect upon various epistemologies, ontologies, and methodologies of geographic analysis of big data and allows us to see and explore the fruitful contexts and meanings of diverse socio-spatial, cultural, political, and technical boundaries of knowledge in the digital world.

Notes

1. Smolan and Erwitt, *Human Face of Big Data*; Kitchin, "Big Data and Human Geography"; Yakabuski, "Big Data Should Inspire," cited in Barnes, "Big Data, Little History."
2. Meeker and Wu, "Internet Trends"; Frisch, "Year of the Zettabyte."
3. Executive Office of the President, *Big Data*.
4. Laney, "3D Data Management"; Kitchin, "Big Data, New Epistemologies"; Goodchild, "Quality of Big (Geo) Data."
5. Zikopoulos et al., *Understanding Big Data*; Executive Office of the President, *Big Data*.
6. Crawford, "Anxieties of Big Data."
7. Manovich, "Trending."
8. Sui and Goodchild, "Convergence of GIS and Social Media."
9. Cope and Elwood, *Qualitative GIS*; Sui and DeLyser, "Crossing the Qualitative-Quantitative Chasm I"; DeLyser and Sui, "Crossing the Qualitative-Quantitative Chasm II."
10. Gorman, "Danger of a Big Data Episteme"; Meeker, "Top Mobile Internet Trends."
11. Many mainstream social media are now spatially linked. For example, Foursquare, a company founded in 2009, provides a location-based service that lets users check in to various locations using their mobile phones. See Foursquare, "About." Twitter has recently launched a tweet-location service feature following this new trend. See Schonfeld, "Twitter Adds Places"; and Sui and Goodchild, "Convergence of GIS and Social Media."
12. Miller, "Social Scientists."
13. Kent and Capello Jr., "Spatial Patterns and Demographic Indicators"; Vieweg et al., "Microblogging."
14. Acar and Muraki, "Twitter for Crisis Communication"; Earle, Bowden, and Guy, "Twitter Earthquake Detection."
15. Ghosh and Guha, "What Are We 'Tweeting?'"
16. Gayo-Avello et al., "Limits of Electoral Predictions."

17. Skoric et al., "Tweets and Votes"; Kim Sang and Bos, "Predicting the 2011 Dutch Senate Election."
18. Tumasjan et al., "Predicting Elections with Twitter"; Kim Sang and Bos, "Predicting the 2011 Dutch Senate Election."
19. Jungherr, Jürgens, and Schoen, "Why the Pirate Party Won"; O'Connor et al., "From Tweets to Polls"; Gayo-Avello et al., "Limits of Electoral Predictions."
20. This also shows how the quantitative analysis part of our research is designed.
21. Batty, "Editorial"; Boyd and Ellison, "Social Network Sites."
22. Tsou et al., "Mapping Social Activities."
23. MacEachren et al., "Geo-Twitter Analytics."
24. Crampton et al., "Beyond the Geotag"; Li, Goodchild, and Xu, "Spatial, Temporal, and Socioeconomic Patterns"; Stefanidis et al., "Demarcating New Boundaries."
25. Hawelka et al., "Geo-Located Twitter as Proxy."
26. Xu, Wong, and Yang, "Evaluating the 'Geographical Awareness'"; Ghosh and Guha, "What Are We 'Tweeting?'"; Kent and Capello Jr., "Spatial Patterns and Demographic Indicators"; Li, Goodchild, and Xu, "Spatial, Temporal, and Socioeconomic Patterns."
27. Brewer and Hunter, *Foundations of Multimethod Research*.
28. Creswell, *Research Design*; Tashakkori and Teddlie, *Handbook of Mixed Methodology*.
29. Barnes, "Retheorizing Economic Geography"; Cope, "Home-Work Links"; Haraway, *Modest_Witness@Second_Millennium*; McLafferty, "Counting for Women"; Poon, "Qualitative Methods"; Plummer and Sheppard, "Emancipatory Economic Geography"; Nightingale, "Feminist in the Forest"; Mattingly and Falconer-Al-Hindi, "Should Women Count?"
30. Cope and Elwood, *Qualitative GIS*; Leszczynski, "Quantitative Limits"; Schuurman and Pratt, "Care of the Subject"; Kwan and Knigge, "Guest Editorial"; E. Sheppard, "Quantitative Geography"; Wilson, "Towards a Genealogy."
31. Miles and Huberman, *Qualitative Data Analysis*; Altheide and Schneider, *Qualitative Media Analysis*.
32. Denzin and Lincoln, *Strategies of Qualitative Inquiry*.
33. Murthy, "Digital Ethnography."
34. Murthy, "Digital Ethnography," 849.
35. Sandercock and Attili, "Digital Ethnography"; Pink, "Visual Ethnography."
36. Creswell, *Research Design*; Johnson and Onwuegbuzie, "Mixed Methods Research"; Tashakkori and Teddlie, *Handbook of Mixed Methodology*.
37. The data set was provided by the DOLLY (Data on Local Life and You) project at the University of Kentucky.

38. Glaser and Strauss, *Discovery of Grounded Theory*; Strauss and Corbin, *Grounded Theory in Practice.*

39. Cope, "Coding Qualitative Data."

40. Glaser and Strauss, *Discovery of Grounded Theory.*

41. This was a moment that we realized the challenges of working with big data, especially conducting qualitative research with large data sets. On average we could read and code about 200 to 250 tweets per hour. It took more than 100 hours to perform initial coding on the 3 percent sample.

42. These codes are exemplary results of the in vivo coding process. We deleted codes that included any profanity or sexually assault references or inappropriate words or expressions from the list of codes for both I-502 and R-74.

43. The sources of these tweets are removed to preserve anonymity.

44. "Twitter ID" refers to the username associated with the tweet.

45. Bailey and Gatrell, *Interactive Spatial Analysis.*

46. Silverman, *Density Estimation Statistics.*

47. U.S. Census Bureau, *2010 U.S. Census.*

48. M. Graham and Shelton, "Geography and the Future of Big Data."

49. Leetaru et al., "Mapping the Global Twitter Heartbeat."

50. Hahmann, Purves, and Burghardt, "Twitter Location (Sometimes) Matters."

51. Postill and Pink, "Social Media Ethnography."

52. See, for example, Eckert, "Tropes 2.0"; and Wilson, "Towards a Genealogy."

53. Zikopoulos et al., *Understanding Big Data.*

54. Crawford, "Anxieties of Big Data."

55. Barnes, "Big Data, Little History."

56. Johnson and Onwuegbuzie, "Mixed Methods Research."

57. Sui and DeLyser, "Crossing the Qualitative-Quantitative Chasm I."

6   Geosocial Footprints and Geoprivacy Concerns

Christopher D. Weidemann, Jennifer N. Swift, and Karen K. Kemp

Social media companies such as Facebook, Google+, and Twitter store large sets of personal information about their users. In the context of social media the information the general public contributes, either voluntarily or involuntarily, on a daily basis is considered big data—petabyte-sized warehouses of personal information collected and maintained over decades by social media platforms.[1] As of July 2015 there were 1.49 billion users on Facebook each month, Google+ had 20 million unique mobile users every month, and Twitter had 100 million active daily users.[2]

Not only do social media companies store and analyze the information provided to them through registration and status updates, but users also grant these companies access to personal data through Internet browser cookies, search history, and even email conversations. Such companies are permitted to store these data because the user often assumes that their personal information will be used only to customize their own experience. Such personal data often include extensive location intelligence, which can be used to track users' interests and activities over time as well as space. This collection of data infringes on users' location privacy, a circumstance known as geospatial privacy, or simply geoprivacy.[3] Geoprivacy is a critical concern in the use of big data by social media companies and also by businesses and local law enforcement and major government agencies that make use of such data.[4]

Social media use has been evolving so rapidly that research into the positive and negative impacts of uses of the personal data generated has been challenged to keep pace.[5] Resources for teaching and learning how to build applications that harvest and analyze social media data are also on the rise.[6] Thus the need to address social media privacy in general and geoprivacy in particular is increasing rapidly.

Described below is one attempt to raise social media users' awareness about the many ways location can be overshared. The free, publicly available tool named Geosocial Footprint (geosocialfootprint.com) is a web-mapping application that aims to inform Twitter users about their personal social media location footprint and the potential risks of social media over-sharing. By highlighting areas of concern and providing dynamic instructions on how to reduce oversharing habits, Geosocial Footprint allows users to view and manage their social media location privacy concerns.

First, however, we briefly examine the range of privacy issues arising from the widespread use of social media before turning to the explosion of social media location data mining and its implications for geospatial privacy. Second we outline the design and development of Geosocial Footprint and then review the results obtained by a Twitter user with a high risk rating. Third we examine and analyze the results of feedback collected over a period of four months via a user survey that confirmed Geosocial Footprint achieved its main goal: to educate and inform social media users of the privacy risks of location oversharing. Fourth we briefly describe the next steps in the development of Geosocial Footprint, including the implementation of geographic entity recognition and machine learning to improve the risk calculation over time. Subsequently we discuss the future of location privacy in social media as it relates to the use of direct and indirect volunteered geographic information (VGI). In conclusion we summarize both the positive and negative effects of sharing personal information using social media platforms, the dynamic nature of the developer landscape of an API for a social media provider, and the need for establishing geoprivacy rights and laws to protect the general public.

The Misuse of Social Media Information

Social media users should be concerned with much more than simply the data they provide to social media platforms. Over the past decade there have been numerous studies on how personal information derived from social media can be gathered, analyzed, and utilized by individuals, organizations, and governments, as well as by the social media companies that collect the data. In the context of privacy concerns there have been many studies on the use of social media information as a screening tool by organizations. For example, a 2012 study by Lini found that 35 percent of hiring managers, across a wide array of domains, had rejected an applicant based purely on information they found online and that the screening of new hires by human resource departments through social media searches rose 38.4 percent between 2008 and 2011.[7] Kluemper et al. found that this trend has continued, thus raising concerns about the reliability, standardization, and validity of social media data as well as fairness in regard to discrimination and privacy.[8] Screening via social media is also used by insurance companies and banks to gather and analyze personal data before making decisions about premiums, claims, and loans.[9] It is important to note that while government agencies might now be able to use data obtained from Facebook, Twitter, and other social media in recruiting and hiring, some still choose not to due to liability concerns.[10]

Social media screening doesn't end with employers, insurance companies, banks, and the government. Many types of crimes have been on the rise due to easy access to personal information and location intelligence obtained from social media websites.[11] Thieves use social media as a tool for gathering intelligence and for picking their victims, locating their whereabouts, and disseminating information to their associates. A notable example was uncovered when police in Brazil arrested a gang that used social media to stalk and kidnap individuals and then to maximize their ransom.[12] The gang, which was linked to the kidnappings of up to nineteen individuals, first found their victims online through their social media posts and evaluated their worth through their tweets and social connections, which described extravagant lifestyles. They then stalked the victims physically by visiting the locations they had geotagged or referenced in text and ultimately used this spatial intelligence to kidnap them. Another serious criminal case occurred in 2012 when a blogger encouraged the murder of police officials using social media postings.[13]

Further examples of criminal activity involving social media data abound. In 2008 U.S. vice presidential candidate Sarah Palin's email account was compromised after a thief was able to gather the necessary intelligence from web searches for information posted in her social media outlets.[14] The National Foundation for Credit Counseling notes that social media mining and subsequent online identity thefts "are ground zero for credit card identity thieves."[15] Other crimes, such as robberies initiated by criminals simply checking Facebook postings for potential victims who are traveling, have become commonplace.[16]

These crimes are made possible through the seemingly innocent sharing of information on social media websites. For example, most Internet users don't stop to think that a mother's maiden name, which is a common secondary security question, can be gathered from social media relationships. It is also common for users to share seemingly harmless photographs of their families or homes. Most significantly, social media users often disclose their real-time location to what they assume are their family and friends. To the ill-intentioned, this is yet more information that could be used to profile the social media user's identity and habits.

Exploiting Social Media Location Intelligence

Many researchers have shown how social media streams can provide considerable amounts of "ambient" information. Humphreys, Gill, and Krishnamurthy have described how phrases extracted from the text of tweets can be used to classify them according to activities, locations, proper names, times of day, and information about the authors.[17] In a similar vein Stefanidis et al., by capturing geospatial information from social media streams, have provided a methodological foundation for mining these streams to extract not only directly geotagged information but also ambient geospatial information, which consists of data that reference a location yet are not spatially located.[18]

In a study that attempted to raise awareness about cybercasing, Friedland and Sommer demonstrated how it was possible to use publicly avail-

Weidemann, Swift, and Kemp

able geospatial information associated with geotagged pictures and videos to infer real-world situational awareness for questionable purposes.[19] They argue that most submitters of the geotagged media are unaware that the location information being published is metadata that is easily extracted and manipulated.

Like all kinds of social media data, location intelligence is often used for good. However, there are many ways in which location data can be misused. Given the ease with which location information can be scraped or collected from the web, several websites have been developed that expose the consequences of such oversharing. There are many examples of websites that simply map the location tags associated with tweets. Tweetping.net (currently operational) builds a global map of all live tweets. What it lacks in micro and privacy focus, it makes up for by displaying geolocated tweets in nearly real time. At the personal scale Tweography .com (no longer functional) could be used to extract and display all of a single user's geolocated tweets on a map.

Weknowwhatyouredoing.com was originally developed as an educational website and described as a social media privacy experiment. Online until 2015, this application focused on ambient information in the entire Twitter stream, not just from individual users. By sorting and filtering tweets based on key phrases that divulged specific personal or identifiable information, the site's main page provided a stream of tweets from current Twitter users who, based on the context of their tweets, were determined to be either hung over, taking drugs, or speaking ill of their boss or who had recently published a new phone number on Twitter.

By exploiting ambient geospatial information, WeKnowYourHouse .com, which was online only briefly in 2012, actively scanned Twitter for geolocated tweets that made a textual reference to a user's home. From there the web application mapped the location using Google Street View and automatically responded to the user on Twitter with a tweet stating his or her address. While all the information the web application used was public, Twitter very quickly responded to privacy concerns and suspended the account, but not before many news outlets had reported on it.[20]

Also in 2012 PleaseRobMe.com made it possible to see the check-in activity of a Foursquare user simply by entering a Twitter username. The results could indicate whether or not that user was away from home, implying that the user's home could easily be burglarized. While the application was simple and provided no location context, like WeKnowYourHouse it served as a reminder to all users that Twitter data, and any location data associated with it, are public. Regardless of what their names imply, these sites actively helped Twitter users stop disclosing personal information by alerting potential targets to their oversharing habits. The social media monitoring company Geofeedia also came online in 2012, offering a location-based, open-source solution geared toward supporting police tracking of individuals using tweets and providing spatial analytics for journalists and businesses.[21]

Like the applications described above, Geosocial Footprint sheds light on the ease with which personal location data can be extracted and analyzed from tweets. This application gives the user a dynamic experience, displaying a user's spatial footprint and providing feedback for decreasing oversharing risks. While the application focuses on Twitter, these same techniques can be adapted to collect and analyze other social media data. This value-added information raised awareness in Twitter users who utilized the application and replied to the survey regarding weaknesses as well as strengths in the privacy options offered.

Having explored the need for social media privacy and the ease with which location intelligence can be extracted for both good and harmful uses, the next section discusses the design of the Geosocial Footprint web application. However, it is important to note first that since the release of Geosocial Footprint in August 2013, UC Berkley's Teaching Privacy group has released a similar application named Ready or Not? (app.teachingprivacy.com) that allows users to visualize their Twitter footprint.[22] It also performs a temporal analysis of the user's tweets, which can show patterns in weekly tweeting habits. This National Science Foundation–sponsored Teaching Privacy project includes an interdisciplinary team of researchers and educators aiming to develop tools designed specifically to inform students in K–12 and higher education about their online privacy choices.[23]

Design of the Geosocial Footprint Application

The main objective of Geosocial Footprint is to educate and inform Twitter users of potential location oversharing. System design requirements included a simple user interface, client side application processing, open source technologies, and web application event tracking. A simple user interface is intended to provide a user-friendly experience for the untrained. Client side logic reduces the overall hosting costs and improves the application's response time by allowing the user's Internet browser to handle the data storage and processing. Building upon and further developing open source technologies not only expands the body of knowledge but also reduces the cost of entry for this application and others like it through open innovation.[24] Selective web application event tracking allows for system quality assurance.[25]

The application relies heavily on the use of third-party APIs to access data and third-party libraries to analyze and display results. Google Maps API version 3.3 was used as the web mapping base, Twitter's API was used to access the tweets, and Bootstrap.js and jQuery.js libraries were both implemented in the web application development. In total ten modules were created to move the data from Twitter and to parse, convert, store, geoprocess, analyze, and display results. The code is executed through a web application that requests a user's Twitter ID as input. This application's flow chart is shown in figure 6-1.

To maintain the application's simplicity, the home page of Geosocial Footprint includes a summary of the web application, instructions for use, and links to news coverage this web application received. Users access Geosocial Footprint by entering the desired Twitter username in the input box and clicking the button labeled "Retrieve Tweets." The application then checks the validity of the Twitter username, and if the username is valid, the main user interface box updates and informs the user that the application is retrieving the most recent two hundred tweets posted by the account.

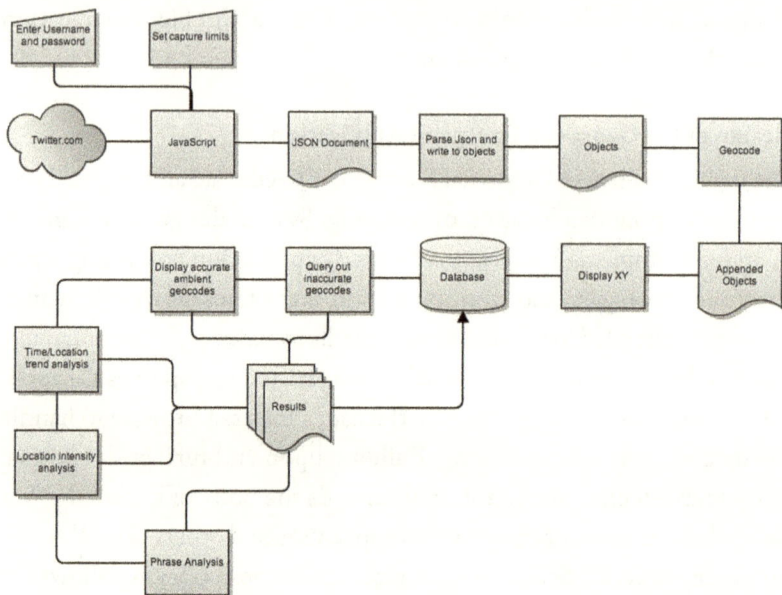

FIG. 6-1. Flowchart of Geosocial Footprint's application components. Created by C. D. Weidemann.

After a valid Twitter username has been entered and tweets have been retrieved, the web application updates the main user interface box with a heat map illustrating the relative density of geocoded tweets. The place-names, latitude, longitude, or other location indicators are provided through Twitter's API as JavaScript Object Notation (JSON). The JSON is parsed during this process, and all locations are extracted for analysis.

Figure 6-2 shows the results for a user who has a high risk rating where an obvious footprint is visible. In the event that the web application cannot geolocate any tweets for the specified user, a text alert stating "Not at Risk" is added to a blank map. Below the map three columns of information show the user's alerts, risks, and suggestions.

User Risk, Alerts, and Suggestions

After a Twitter username has been validated, tweets have been retrieved, and a map has been generated, the user is provided with dynamic alerts, a

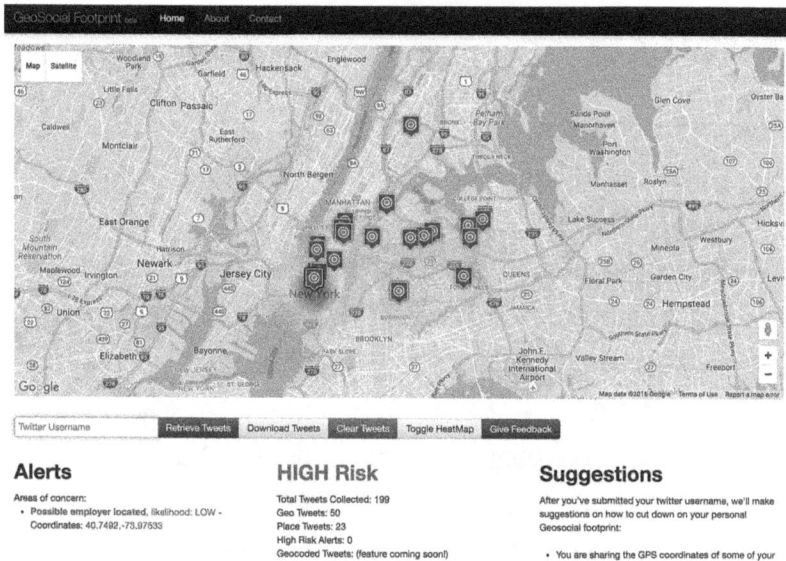

Alerts

Areas of concern:
- Possible employer located, likelihood: LOW -
  Coordinates: 40.7492,-73.97533

HIGH Risk

Total Tweets Collected: 199
Geo Tweets: 50
Place Tweets: 23
High Risk Alerts: 0
Geocoded Tweets: (feature coming soon!)

Suggestions

After you've submitted your twitter username, we'll make
suggestions on how to cut down on your personal
Geosocial footprint:

- You are sharing the GPS coordinates of some of your

FIG. 6-2. Map results of a high-risk Twitter user. Created by C. D. Weidemann from Geosocial Footprint.

location oversharing risk assessment, and suggestions on decreasing said risks. Dynamic alerts use a stored dictionary of key location-identifying phrases for word context matching analysis. This makes it possible to detect tweets that provide a textual clue about what the collected GPS location may be. For instance, the phrase "I'm at home" is in that dictionary. If someone tweeted "I'm at home watching the game" and included a GPS coordinate, the system would alert the user of a potential risk. The phrases have attributed confidence levels, so a term such as "at home" is also in the dictionary but carries a lesser confidence. This is displayed on the application under "Alerts." Alerts in red carry the highest confidence. Figure 6-3 shows an example of the alerts for the user's map shown in figure 6-2.

The alerts section displays the results of the word context matching. These alerts are separated into three categories that represent the likelihood of alert accuracy: low, medium, and high. This likelihood of alert accuracy is based on the word context matching analysis. A classic example

Alerts

Areas of concern:

- **Possible employer located**, likelihood: MEDIUM - Coordinates: 42.25062,-73.78834
- **Possible residence located**, likelihood: LOW - Coordinates: 42.25062,-73.78834
- **Possible employer located**, likelihood: LOW - Coordinates: 40.69893373,-73.97214549
- **Possible employer located**, likelihood: LOW - Coordinates: 40.69903,-73.9556999
- **Possible residence located**, likelihood: LOW - Coordinates: 40.69903,-73.9556999
- **Possible employer located**, likelihood: LOW - Coordinates: 40.68226,-73.92907
- **Possible employer located**, likelihood: LOW - Coordinates: 42.2512,-73.7851
- **Possible employer located**, likelihood: LOW - Coordinates: 42.246784,-73.783241
- **Possible employer located**, likelihood: LOW - Coordinates: 41.0389099,-73.6959229
- **Possible employer located**, likelihood: LOW - Coordinates: 42.2552681,-73.7945099

🐦 Tweet | G+1 | 51

FIG. 6-3. Alert results for a high-risk user. Created by C. D. Weidemann from Geosocial Footprint.

of word context matching is looking for the phrase "I am at home," which is generally used by Twitter users when they are physically tweeting from their home address. When the occurrence of known words or phrases that divulge additional location information increases within a tweet, so does the likelihood rating. The likelihood category is not only displayed as text, but the color of the text for each alert changes based on the alert level. To increase the usability of the alerts, when the user clicks on an alert, the map centers and zooms to the selected alert.

The risk of oversharing is displayed in the center column below the map. The risk assessment is based on the sum of the total number of place tweets (tweets containing place-names) and geolocated tweets (tweets with location information automatically attached) divided by the total number of observed tweets. To account for the higher risk of tweets observed during the alerts analysis, the percentage of high-risk alerts (number of high-risk alerts divided by total number of observed tweets) is also added to the risk assessment. The risk ratings were derived by taking a large sample of users and calculating their Geosocial Footprint risk. As a proxy for long-term risk assessment, the top 20 percent of the users were classified as high risk and the derived classification was used for the initial application development. The heading of the risk column changes according to the resulting risk level: low, medium, and high. This text also changes color based on the risk level. Below the risk rating the user is also provided with a summary of what was collected for the specified Twitter user, including the collected number of total tweets, geolocated tweets, place tweets, high-risk alerts, and geocoded tweets.

The suggestions shown in the right-hand column in figure 6-2 provide users with hints on how to decrease their personal oversharing risk. These suggestions are dynamically presented based on the users' alerts, risk rating, and the types of location tweets they disclose. For instance, each user who has a GPS-enabled tweet is informed and shown a link to a web page that provides information intended to educate users about their geoprivacy. Users who are alerted to tweets that may expose the location of their home are provided with a link to a USA Today article that discusses the dangers of social oversharing.[26] The suggestions were originally compiled in late 2013 from a list of resources, links, and videos on social media oversharing.[27] These suggestions are updated annually.

Finally, after a user's tweets have successfully been retrieved, all geolocated tweets are available for download as a comma-separated values (CSV) file. The CSV file contains the tweet time, latitude, longitude, and text.

Evaluation of User Experience

During its initial implementation period of August 2013 to January 2014 Geosocial Footprint was evaluated for its efficacy in educating a user on the potential risks of oversharing. User feedback was collected through a survey whose results provided metrics to gauge the educational value and effectiveness of the application. After two minutes of use the web application triggered a pop-up window that prompted the user to participate in an anonymous user survey. The user survey provided an opportunity for feedback in five areas that measure educational value.

If users agreed to participate in the survey, they were forwarded to a Google Forms page and asked to rate each of the following questions:

- "Prior to using Geosocial Footprint, how would you have rated your location privacy on Twitter.com?"
  Possible answer: a scale from 1 to 10, with 1 representing "private" and 10 representing "insecure"
- "After using Geosocial Footprint, how do you rate your location privacy on Twitter.com?"
  Possible answer: a scale from 1 to 10, with 1 representing "private" and 10 representing "insecure"
- "How would you rate the personalized suggestions you received on reducing your oversharing risk?"
  Possible answer: a scale from 1 to 10, with 1 representing "no help" and 10 representing "extremely helpful"
- "Did you find Geosocial Footprint educational?"
  Possible answer: "yes," "no," and "undecided"
- "If Geosocial Footprint located areas of concern, were any of them accurate?"

Weidemann, Swift, and Kemp

Possible answer: "yes," "no," "undecided," and "Geosocial Footprint did not locate areas of concern for my account."

Finally, users were invited to enter general comments in a blank text box: "We're interested to hear what you have to say. Please share any general comments, concerns, or feature requests."

Results of the User Survey

Analysis of the 352 surveys collected over a four-month period between August 2013 and January 2014 produced useful insights. In response to the first question only 16 percent of the anonymous users reported an insecurity rating of 8 or higher, with 10 being the most insecure, with respect to their feelings about using Twitter prior to using the application, while 50 percent gave a rating of 3 or lower. Figure 6-4 shows the summary of results for the first question.

The second question asked about the effect of using Geosocial Footprint. Results are shown in figure 6-5. After using Geosocial Footprint, roughly half (49 percent) of users who responded to the survey indicated that they had strong confidence (ratings of 3 and under) in their privacy after using the application. However, the number of users who recorded an insecure rating of 8 or above doubled compared to responses to the first question: 30 percent of the users as opposed to only 16 percent for the previous question. This shows that among users who took the survey, Geosocial Footprint did help increase awareness of social media oversharing insecurities. In addition the number of private (rating of 1) users increased from 22 to 30 percent. This shows that the tool not only highlights insecurities but is also able to instill confidence in some users about their privacy.

Figure 6-6 shows a summary of the results from the third question, which asked if the personalized suggestions were helpful. The bimodal distribution of the graph shows that users found the personalized suggestions either extremely helpful or of no help at all. It is important to note that many users who reported that the suggestions were of no help also left comments suggesting their Twitter username didn't produce any geolocated tweets and therefore no personalized suggestions were provided.

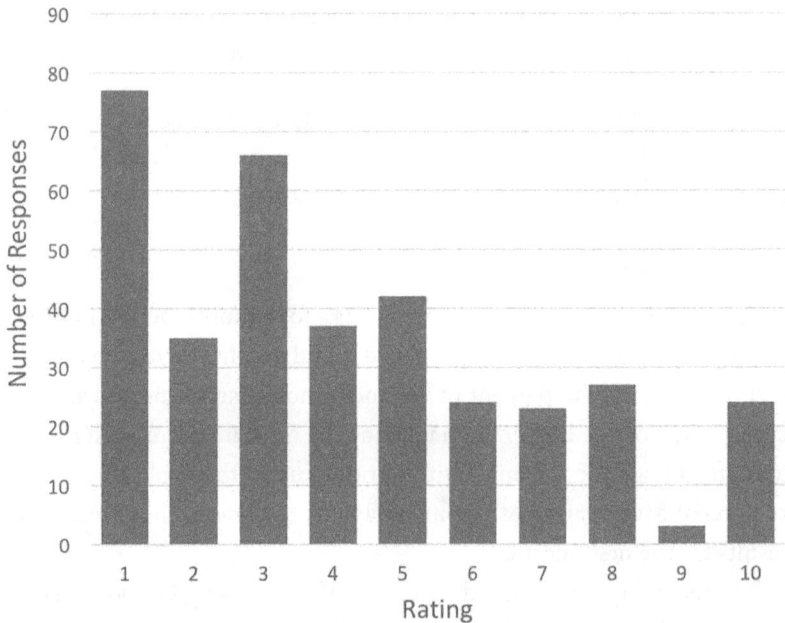

FIG. 6-4. Summary of responses to question 1: "Prior to using Geosocial Footprint, how would you have rated your location privacy on Twitter.com?" Created by C. D. Weidemann.

The fourth question asked if users found the application educational. These responses provide the most convincing results in favor of Geosocial Footprint's effectiveness in educating social media users about location privacy. An overwhelming 82 percent found Geosocial Footprint to be educational, with only 8 percent not finding it educational, while an additional 10 percent of users remained undecided. Responses to this question about educational value are shown in the top portion of figure 6-7.

Finally, the fifth question asked if any areas of concern that were identified were accurate. For this question 54 percent of the users did not have any areas of concern identified. Of the remaining users more than half found that the areas of concern that were identified were accurate. This is strong evidence that word context matching is indeed a viable solution for collected metadata about location-enabled tweets.

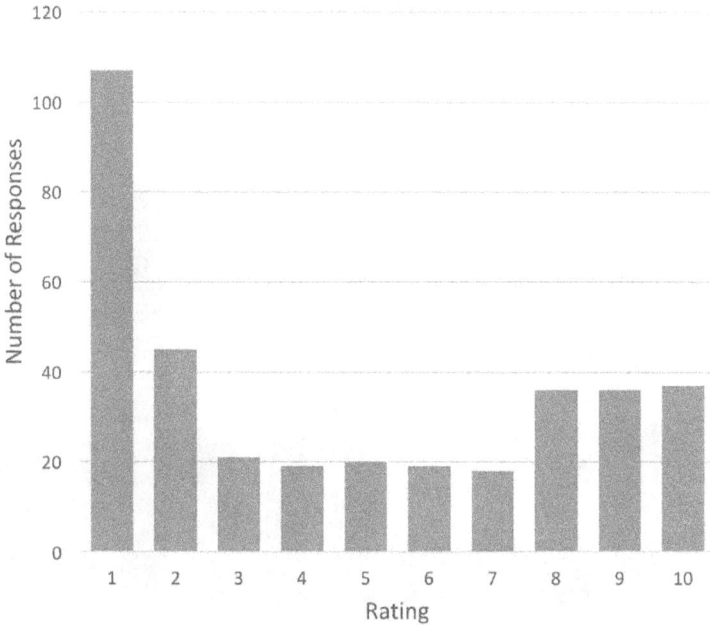

FIG. 6-5. Summary of responses to question 2: "After using Geosocial Footprint, how would you have rated your location privacy on Twitter.com?" Created by C. D. Weidemann.

These combined results provide evidence that Geosocial Footprint was successful in achieving the main goal of educating and informing social media users about the privacy risks of location oversharing. Finally, the personalized suggestions on how to increase social media privacy were deemed successful for users who have geolocated tweets.

Next Steps

A new geocoding tool is planned for a second version of Geosocial Footprint, and it will allow for geographic entity recognition and geocoding. Twitter currently handles the geocoding of place tweets. The new geocoding tool will focus on geocoding the additional text found within a tweet using natural language processing. Entity recognition extracts usable information from strings of text. Extracted geographic entities

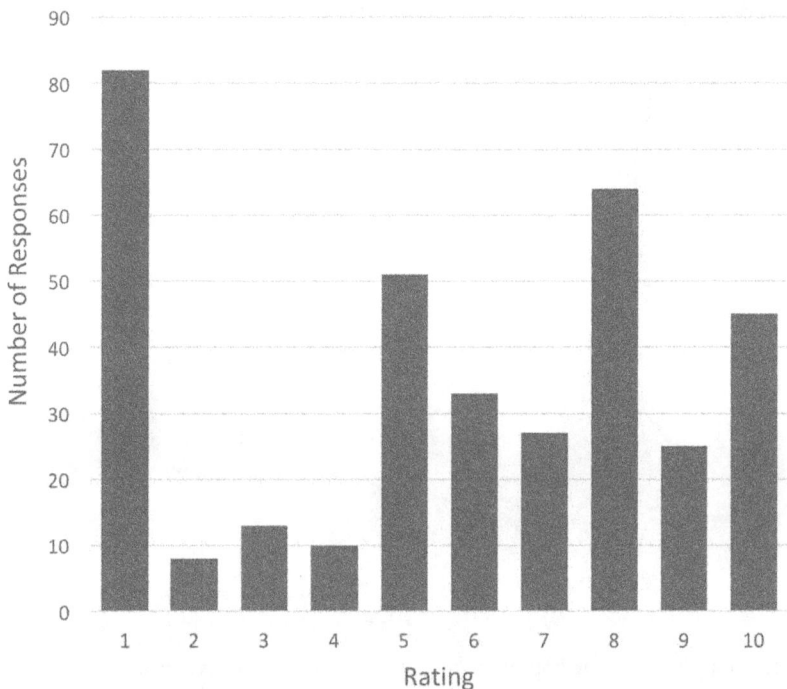

FIG. 6-6. Summary of responses to question 3: "How would you rate the person-alized suggestions you received on reducing your oversharing risk?" Created by C. D. Weidemann.

can be looked up in gazetteers and the relevant geographic coordinates can be assigned to the string of text. Weidemann and Swift have shown that geocoding entire tweets through entity recognition can increase the volume of geolocated tweets, from only 3 percent of the Twitter API firehose stream to nearly 30 percent.[28] Such geocoding would increase the number of geographic coordinates visualized on the map and ultimately help shape a user's Geosocial Footprint to an improved degree.

Machine learning could improve upon the analytics even further by refining the data processing and analytical methods based on the continued use of the application and the results found therein. For example, the risk assessment method could dynamically respond to the continued

Weidemann, Swift, and Kemp

Did you find Geosocial Footprint Educational?

- Yes
- No
- Undecided

36
29
293

If Geosocial Footprint located areas of concern, were any of them accurate?

- Yes
- No
- Undecided
- Geosocial Footprint did not locate areas of concern for my account

93
195
26
44

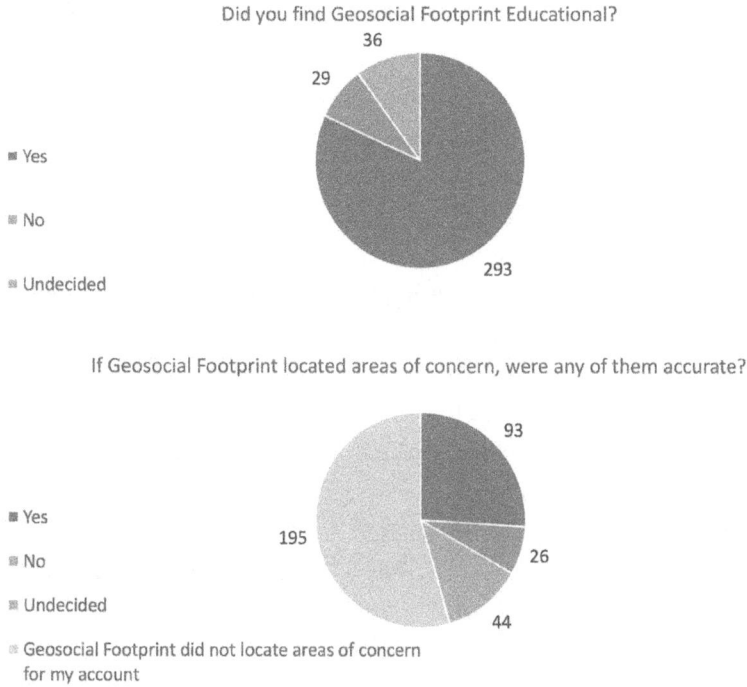

FIG. 6-7. Summary of responses to questions 4 and 5. Created by C. D. Weidemann.

use of the web application and adjust the risk classifications based on statistical findings such as a decrease in the mean percentage of geolocated tweets. An alternative would be for entity recognition and gazetteers to be modified in the geocoding process as users verify the accuracy of such findings.

The Future of Location Privacy in Social Media

Technology changes faster than most users can adapt. In the context of social media the change has brought on many privacy concerns, a small part of which Geosocial Footprint successfully addresses. While sometimes this rapid change brings about negative consequences, many times technology evolves to improve society. Social media has seen negative change in the past, but it is now entering a new era of improvement.

Social media companies, including Twitter, have made initiatives to limit accessibility to user data through the use of authenticated APIs.[29] Other social media providers have tried to better manage data-sharing options and security. Through trial and error social media providers will find solutions that permit rich immersive social media content without compromising the user's privacy. Applications that expand new technology and highlight weakness are necessary in the overall balance of achieving social media privacy. In the context of location privacy specifically a slew of applications will emerge that either harness the locational power of social media platforms to provide innovative uses or that, on the flip side, exploit the weaknesses of such networks.

In the future particular emphasis in the development of social media information harvesting and analysis tools will be on the extraction of direct and indirect VGI.[30] With indirect VGI the user anonymously and autonomously contributes to a geographic information collection through active sensors. OpenStreetMap is a major example of direct VGI. Users consciously contribute to an obvious geospatial system. WAZE, a road navigation and traffic application that becomes more accurate in its depiction of the transportation network as more users contribute to the geospatial system, is somewhere between indirect and direct VGI.[31] While none of these VGI examples use Twitter as their communication network, the same information can be transferred through social media protocols.[32] With third-party APIs such as Twitter to collect and communicate VGI, the development time and effort required to generate new VGI applications can be greatly reduced, permitting more successful implementations.

Many examples of beneficial uses of indirect VGI sources that use social media as a sensor are already in use today in disaster management.[33] For example, Weiler, Grossniklaus, and Scholl utilized Twitter streams to conduct case studies of emotions during recent large public events in Boston, Washington, and during the 2014 Super Bowl.[34] The Federal Emergency Management Agency (FEMA) uses social media event tracking to collect

Weidemann, Swift, and Kemp

indirect VGI to inform decision makers in the event of an emergency.[35] Twitter users do not actively communicate on Twitter for the purpose of informing FEMA's VGI collection, but instead the FEMA application actively listens for geospatial content related to their interests. This same approach will be applied to many different media in the future. Effective privacy constraints need to be in place to reduce the risk of nefarious VGI scraping or data collecting systems while still empowering well-meaning entities like FEMA to save lives.

Since many people do not seem to comprehend the potential negative consequences of sharing location data, some argue that users must be given the opportunity to set granular and custom location data sharing preferences.[36] These location privacy preference settings might involve methods that allow a user to select the spatial resolution and accuracy for the location data they are required to provide. Such settings would allow system designers to take responsibility for protecting users' privacy through location-anonymizing algorithms. Thus if an application's spatial resolution or accuracy requirements do not meet the location privacy preferences of a user, then anonymization techniques that prevent the disclosure of the user's location should be initiated to limit the geoprivacy risks exposed through the application.[37] Anonymization techniques could also be used to protect the public from geofencing where global positioning systems (GPS) or radio frequency identification (RFID) data are gathered from the public's use of these technologies; such techniques would set a virtual perimeter around a location for crowd control or other security reasons.[38]

Another important consideration when designing web applications that utilize VGI is the dynamic nature of social media development communities. Social media companies such as Twitter are beginning to realize the monetary value of the data at hand and can at any time choose to restrict or withdraw free use of their development APIs. For example, in 2015 Twitter announced a plan to move much of its popular free JSON development API environment to their subsidiary company, Gnip, that specializes in business-to-business data aggregation.[39] After this move some functionality would be monetized.[40] This change highlights the highly dynamic Twitter application development landscape, where the privacy concerns of users, business and government use of Twitter feeds, and developer support are constantly moving targets. Fortunately as of October 2015 the publicly reported impending changes to Twitter's development APIs will not affect the functionality of the Geosocial Footprint application.

There is also a need for establishing geoprivacy rights and laws.[41] Privacy laws need to be updated to reflect evolving social media trends. For example, who can or should protect the general public from malicious use of personal data harvested from social media platforms? Some government entities are in the process of stepping up regulations concerning sharing of social media data. Recently the European Court of Justice declared that under the 2000 "safe harbor" agreement Facebook could no longer send personal data on European citizens to U.S. servers that are deemed potentially unprotected from government spying.[42] Although the intent is to protect European citizens' private information, the rejection of the agreement could slow data traffic and thus impede government and business activities around the world.[43] It is anticipated that this ruling will have a serious negative impact on the economies of Europe and the United States. The privacy rights of citizens and the financial constancy of nations may hang in the balance.

Geoprivacy is continuously evolving as consumers learn more about their online presence, social media companies better understand the value and potential uses of their data, and researchers investigate the long-term impact of social media on science, culture, crime, business, and government. The future holds many opportunities not only to advance the use of geosocial data but also to continue to educate and inform the general public about their social media location privacy. As the results of this study show, simple visualization in the form of a map is an effective tool for educating users about oversharing. It is also clear that social media outlets should perform user outreach and allow users to proactively monitor their social media footprint within their social media platforms. This is the responsible path forward for all concerned.

Weidemann, Swift, and Kemp

Notes

1. Lee, *Facebook Nation*; M. Smith et al., "Big Data Privacy Issues"; Thusoo et al., "Data Warehousing"; Debatin et al., "Facebook and Online Privacy."
2. C. Smith, "Geospatial Encountering."
3. Nittel et al., "Emerging Technological Trends"; Nurse, "Exploring Risks"; Kounadi and Leitner, "Why Does Geoprivacy Matter"; boyd and Crawford, "Six Provocations for Big Data"; Shanley et al., "Tweeting Up a Storm"; Monmonier, "Spying with Maps"; Dobson and Fisher, "Geoslavery."
4. I. Brown, "Social Media Surveillance"; Shekhar, Feiner, and Aref, "From GPS and Virtual Globes"; Executive Office of the President, *Big Data*.
5. Punathambekar and Kavada, "Debating Big Data."
6. Gambs, Killijian, and Cortez, "Show Me How You Move"; Volkova et al., "Social Media Location Intelligence."
7. Lini, "Employers Eyeing Twitter."
8. Kluemper et al., "Social Networking Websites."
9. Romero, "Are Insurance Companies Spying"; Roulin and Bangerter, "Social Networking Websites."
10. Tufts, Jacobson, and Stevens, "Status Update Social Media."
11. CBS News, "Social Media Related Crimes."
12. Fox News World, "Brazil Police Bust Kidnappers."
13. Leibowitz, "Philadelphia Blogger Charged."
14. Jaeger, Paquette, and Simmons, "Information Policy."
15. Benda, "Sharing Information on Social Media."
16. Cheng, "Facebook Chat."
17. Humphreys, Gill, and Krishnamurthy, "How Much Is Too Much?"
18. Stefanidis, Crooks, and Radzikowski, "Harvesting Ambient Geospatial Information."
19. Friedland and Sommer, "Cybercasing the Joint."
20. Perez, "Following Twitter Suspension."
21. Geofeedia, *Case Study*.
22. Friedland, "Privacy Concerns."
23. Bernd et al., "Teaching Privacy."
24. Huizingh, "Open Innovation."
25. Nitze, "Evaluation of JavaScript."
26. Shelly, "Social Media."
27. Shelly, "Social Media"; Groeneveld, Borsboom, and van Amstel, "Over-Sharing and Location Awareness."
28. Weidemann and Swift, "Social Media Location Intelligence."

29. Kepes, "How to Kill Your Ecosystem."
30. Haworth, Bruce, and Middleton, "Emerging Technologies"; Weiler, Grossniklaus, and Scholl, "Situation Monitoring"; Xiao, Huang, and Wu, "Understanding Social Media Data"; Fugate, "Understanding the Power."
31. C. Smith, "Geospatial Encountering."
32. M. de Oliveira et al., "Leveraging VGI."
33. Saleem, Yisha, and Ruths, "Effects of Disaster Characteristics"; Xiao, Huang, and Wu, "Understanding Social Media Data."
34. Weiler, Grossniklaus, and Scholl, "Situation Monitoring."
35. Fugate, "Understanding the Power."
36. Bettini and Riboni, "Privacy Protection"; Krumm, "Survey of Computational Privacy."
37. Bettini and Riboni "Privacy Protection"; Nurse, "Exploring Risks"; I. Brown, "Social Media Surveillance."
38. Bohli et al., "PrivLoc"; R. Oliveira et al., "Intelligent Model."
39. Forbes, "How to Kill Your Ecosystem"; TechCrunch, "Twitter Cuts Off DataSift."
40. M. Anderson, "Twitter's Withdrawal."
41. Nittel et al., "Emerging Technological Trends"; Shekhar, Feiner, and Aref, "From GPS and Virtual Globes"; Executive Office of the President, *Big Data*; Loeffler, "Privacy Issues."
42. CTV News, "Why Europe's Facebook Users Have Cause."
43. U.S. News, "How the Safe Harbor Law Threatens."

7 Foursquare in the City of Fountains

Using Kansas City as a Case Study for Combining Demographic and Social Media Data

Emily Fekete

In light of the cultural turn in economic geography emerging since the early 2000s, the subdiscipline has been opened to new research topics.[1] One of the many areas where economic geography has begun to take culture more seriously is the study of consumption. Consumption, largely regarded as a highly individual activity, is based on the many choices people make in their daily lives and on social and cultural factors. Because of the individual nature of consumption, it is often difficult to study in a broader sense.[2] With the recent growth of geographic information systems analysis, research in geodemographics has reemerged but has largely existed outside of academe in the United States and as a way to profile consumers for marketing purposes.[3] With an increase in research using big data and the geoweb, new data sources have become available to academic researchers. Social media networks such as Yelp, Urban Spoon, Google Places, and Foursquare can provide information on individual consumption decisions on a larger scale.[4]

Despite the availability of these data sources, it is questionable as to whether or not they can be relied on to provide the best interpretation of consumption trends. Many online network sources of big data tend to be user generated. As a result the data are often not representative but rather a reflection of greater societal biases and the historical development of specific physical locations. Therefore the coding of the application itself is suspect, as it is a result of the partiality of the coder and those who use the specific Internet-based site. This chapter presents a case study on data from the social network Foursquare in Kansas City, Missouri, to examine several issues that potentially arise from an overreliance on big data. Researchers must recognize that online big data is more often than not a mirror image of offline social concerns. Big data studies can be helpful in illuminating those on-the-ground issues that warrant future study requiring different methodological foci.

Foursquare: A Geographic Social Network

Foursquare is a location-based application created for smartphone users. Originally designed for users to check in to a location and share their location information with their friends, Foursquare has amassed an extensive database of digital establishments for on-the-ground sites of consumption. Using a combination of the GPS on a smartphone as well as nearby wireless signals, Foursquare triangulates a user's location to provide a list of physical places where that person could be (fig. 7-1). Users then select the store, restaurant, or other establishment corresponding to their actual location and broadcast that information to their friend network.[5]

Foursquare was originally launched in 2009, and by mid-2015 it had grown to a sixty-million-user base worldwide.[6] In order for the company to amass such a large database of digital locations, it relied heavily on its users to create the check-in points through gamification elements. Users gained points and competed against their friends for the most check-ins as well as the creation of new establishments. In 2014 Foursquare deemed its location database expansive enough to move on to the next phase of the company—a "focus solely on exploration and discovery, finally positioning itself as a true Yelp-killer in the battle to provide great local search."[7] Currently the gamification features have been removed from the Foursquare application, which now has as its focus searching for consumer establishments in a person's immediate vicinity. A separate application, Swarm, has been created for those who still wish to play the check-in game. Despite the recent changes within the Foursquare company in terms of the functionality of the application and the company's goals, Foursquare's database provides extensive information about where people are going to consume goods and services.[8] Many of the places where people check in through the application are retail establishments or restaurants, and they often do so because these businesses will offer their customers discounts or free items for checking in.

In addition to sharing location information with a friend network, Foursquare also has a feature called Explore that is similar to other services such as Yelp and Urban Spoon. It allows users to search for a specific type of good or service based on their current location and on nearby businesses that are in the Foursquare database. While Explore is marketed as a useful feature when on vacation or in a new place, it is also an example of a way to filter the world and remove users from distraction to focus on something specific.[9] However, filters such as Explore that are written into the code of a program itself have the potential to become problematic as they reflect a larger societal bias through their coding and, therefore, in the creation of the data.[10]

Social biases reflected through code have been present in other user-generated web applications.[11] For example, Stephens analyzed the gender bias in the creation of content in OpenStreetMap and Google Maps. In OpenStreetMap Stephens finds a wholly gendered space where there are several categories for places that commodify women (e.g., strip clubs, escort services) but where suggested categories for traditionally feminized spaces of care (e.g., child care, hospices) have been routinely rejected as necessary additions to the mapping platform by other users.[12] Excluding some businesses becomes a problem, because many applications, including Apple products, have stopped paying Google for their mapping services in favor of OpenStreetMap.[13] Google Maps is unfortunately no more gender friendly than OpenStreetMap, as "men serve as the gatekeepers of local knowledge and determine what is significant enough to be represented on the base map."[14] Both Internet mapping platforms reflect a male view of the landscape, bringing to mind Rose's consideration of the "male gaze."[15]

Race has also been reflected upon with regard to the geoweb. Crutcher and Zook addressed the posting of images of flood damage to Google Maps and Google Earth in New Orleans following Hurricane Katrina.[16] Although the flooding hit areas in the affluent white neighborhood of Lakewood as hard as it did areas in the lower-income, predominantly

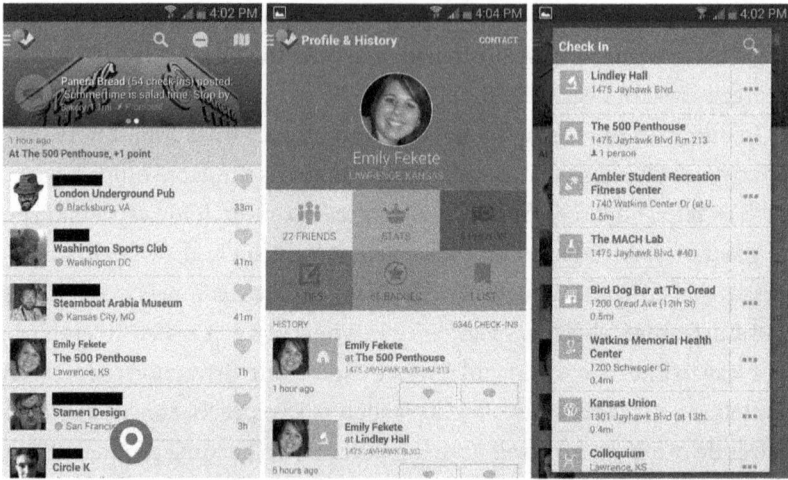

FIG. 7-1. *From left to right*: Foursquare homepage showing locations of friends; Foursquare profile page; Foursquare check-in screen showing nearby locations. Courtesy of Emily Fekete.

African American Ninth Ward, images and comments posted to Scipionus and linked to Google Maps and Google Earth were largely of Lakewood rather than the Ninth Ward. The difference in the number of images posted online between these two neighborhoods is similar to the number of user-generated placemarks on Google Maps: there are more placemarks in white neighborhoods in New Orleans than in African American areas. Crutcher and Zook conclude that "while Google Earth did not create the economic and racial divides present in society, the use of its mapping services in the post-Katrina context reflect[s] this gulf and [has] arguably reinforced and recreated racialized cyberscapes."[17] The presence of a racially divided geoweb questions the assumption that the Internet is a democratic space, representative of all people, just as Stephens noted with regard to gender.[18] The geoweb reflects rather than eradicates social discrepancies on the ground.

While Foursquare is an excellent resource for observing consumption trends in urban locations, the data source is not perfect. Because users generate the venues, the information and the exact locations may not be completely accurate or they may reflect the values of the users themselves. There may also be places that do not have a corresponding Foursquare venue. However, because Foursquare users are an active community and shape the data themselves, Foursquare has the potential to be a rich and detailed database. Similar to other forms of crowdsourced data, the Foursquare database is constantly being updated and policed by users who want an accurate and useful space to share with their friends.

It is important to note that the data are also limited to people who use Foursquare and have a smartphone. While it may provide a good sample of the population on which to draw impressions of consumer patterns, it is inevitably biased. According to a Pew study on cell phone usage in the United States, about 77 percent of adults own a smartphone and about 90 percent of smartphone users have used their smartphone to obtain information related to their location, such as mapping applications or location-based services including Foursquare.[19] This number is heavily skewed by age, education, and income. While smartphone ownership is increasing, it is still most prevalent among younger generations, those with higher household income, and those with higher education levels.[20] Although Foursquare places are the focus of this chapter, it is important to understand the user base that is creating Foursquare venues and checking in to these locations, thus determining the popularity of each place. Places are dynamic and dependent upon their relationships with people, things, and other places. The users who are generating Foursquare places and checking in to these locations are shaping the nature of the online venue and therefore having an effect on offline action. Foursquare places are a reflection of the Foursquare user base in that the existence and popularity of these places is driven by the network of users. Therefore, while Foursquare does have the potential to provide extensive information about offline retail spaces, it is, as all places are, a product of its relationship with those who actively use the service.

Data Extraction and Visualization

To access its venue data Foursquare permits the public to use its application programming interface (API) for free via its website.[21] The Foursquare

API allows users to request data points from the area within a specified distance from a latitude and longitude point. For this chapter data were collected from Foursquare between August 20 and 26, 2013, and only those points categorized as "shop and service" or "food" were used in order to look specifically at sites of consumption. Foursquare categorizes each location point in several primary categories and an extensive variety of subcategories. The categories "shop and service" or "food" were chosen because they best represent those types of places people would go to consume goods and services. It is likely that people are more willing to check in to consumption-based sites because of the way the application is used by its user base.[22] Users want to be able to meet up with friends to socialize (e.g., at a bar or restaurant) or are enticed by special promotions offered by businesses. These primary categories include retail establishments, restaurants, and grocery stores, as well as services such as fitness centers, hairdressers, or pet day-care facilities.

While visualization should not be the only way information from Foursquare is analyzed and understood, the sheer size of the data sets makes it difficult to appreciate the social network from any other viewpoint. By creating maps of the Foursquare venues found in a variety of cities and linking them to socioeconomic characteristics, one can discern patterns that can be examined in detail to get a richer comprehension of consumption and individual economic choices in these locations. This chapter specifically addresses how a user-generated social network can provide data to address patterns of consumption on the ground. Looking at consumption patterns from a big data perspective can help illuminate other social issues to consider in more detail using other methodological approaches.

The data on shop-and-service venues and food venues were mapped proportionally based on the number of visitors each location had received to date. The symbology for the venues is a series of white and dark gray circles displayed at 40 percent transparency to show overlap in venues. Some of the data points will therefore appear light gray in color. The maps are included in this chapter to provide a visual representation of the Foursquare venue locations. Because this chapter is looking at places, it is important to note not just the popularity of locations but also their distribution and location relative to census tracts and other areas of Kansas City, Missouri. By mapping the Foursquare locations, a better illustration of where the venues are located in relation to various demographics is provided, thus illuminating gaps in Foursquare service within Kansas City. Connecting consumption data from the Foursquare network to demographic information has the potential to point out areas of a city or region that may be underserved in terms of the types of goods and services available. These areas can be ground-truthed to see what types of economic development would be most beneficial for the future vitality of various census tracts.

A Pearson's correlation was run to determine if there was a significant relationship between the number of Foursquare venues in a census tract and the demographic information for that tract. The number of Foursquare venues was correlated with the following variables: total population, ethnic/racial profile (white, African American, Hispanic), median family income, median age, and percentage of the population with a bachelor's degree or higher. These correlations were then tested for statistical significance. Also uploaded to ArcMap is information from the U.S. Census Bureau at the census tract level on median family income, percentage of whites, percentage of African Americans, percentage of Hispanics, and median age. The census data used are the five-year estimates from the American Community Survey in 2012.[23] Census tract data were used because they have been shown to be an accurate measure useful for understanding larger-scale neighborhood dynamics.[24] Only the Foursquare locations in the central city were used because of the high volume of retail establishments in the area and the associated difficulty with collecting all data points. Not including data from surrounding suburbs may present a bias in the results because it is likely that those people consuming goods and services in the inner city are from the surrounding metro area. Similarly people are mobile and are able to consume in areas that differ demographically from their own areas. Therefore the consuming populations in each census tract and businesses in each tract may not exactly align.

Kansas City as a Case in Cyberspace

Kansas City, located on the state line between Kansas and Missouri, has a history of redefining itself economically to maintain its presence as an important urban center in the United States. Embedded in the context of Kansas City is an account of ethnic and economic segregation that can still be seen today by looking at census tract data and Foursquare venue locations. As a new technology hub in the United States, Kansas City is positioned well for engaging with hybrid spaces of the physical and online worlds. Kansas City, with a population of 464,310, is also fairly demographically diverse. According to the 2010 census the median age of city residents was 34.6 years, median income was $45,150, and the population had an ethnic profile of 54.9 percent white, 29.9 percent African American, and 10 percent Hispanic. Almost 31 percent of the population had a bachelor's degree or higher.[25]

Kansas City, Missouri, has deep historic roots in the rural central United States. Being located close to Independence, Missouri, Kansas City was founded by early fur traders moving through the state and trading along the Missouri River.[26] Despite the lack of mineral resources for early industrialization found in places such as Pittsburgh, Pennsylvania, Kansas City's proximity to farm and ranch lands allowed for the urban area to grow as a center of meatpacking and milling.[27] The city has maintained its position as a transportation hub, changing from a river trading town to a railroad terminus to a center for trucking and wholesale distribution.[28]

Having experienced industrial growth somewhat later than other cities in the United States, Kansas City did not attract immigrant diversity until much later in its history. After Reconstruction a number of African Americans moved northward and settled in the relatively unsegregated metropolis.[29] During the 1930s and 1940s, however, the city made the decision to concentrate public housing in one inner-city location, and as a result many African American families settled on the eastern side of the city, solidifying a distinctive African American core of the downtown.[30] The most distinctive elements of Kansas City culture—jazz and barbecue—were products of these neighborhoods.[31] Recently there has been a sizable Hispanic immigrant population that has begun to settle in Kansas City, though this population is somewhat spatially dispersed.

The architectural work of J. C. Nichols, notably the Country Club Plaza and Country Club District (the largest contiguous master-planned urban area in the United States), aided in subdividing the city in terms of economic means by creating desirable racially segregated neighborhoods for wealthy families along the Kansas-Missouri state line.[32] The work of Nichols and others gave rise to nicknames for the city, such as Paris of the Plains and the City of Fountains, intended to help residents combat the notion of living in a backwoods, rural society.[33]

Recently Kansas City has become a center for technology with the development of the Sprint Campus and World Headquarters in 1997, as well as the decision made by Google to use Kansas City as a test site for its new Google Fiber network. The success of implementing the high-speed fiber network in Kansas City led Google to publicize its increased availability in thirty-four other locations across the country. With the expansion of Kansas City enterprise to include several corporations focused on technology, it seems plausible that social media applications like Foursquare would be widely known and used.

Kansas City boasts a relatively high percentage of businesses represented on Foursquare. With 1,558 retail venues and 1,110 accommodations and food venues (a total of 2,668) listed in the economic census and 835 Foursquare food and shop-and-service locations, approximately 31 percent of establishments in Kansas City have online counterparts on Foursquare. As shown by figure 7-2, the majority of the Foursquare venues in Kansas City are located along the western edge of the city on the boundary with Kansas City, Kansas. One explanation for this trend could be the historical growth of the city and the push by J. C. Nichols to develop shopping plazas in the southwest corner of the city. The venues here also align with demographic characteristics, as demonstrated later, since this area contains the largest percentage of white, high-income, and older populations. Other hot spots for venue locations are along major highways or at highway intersections.

TABLE 7-1. Most popular sites of consumption in Kansas City, Missouri

TOP TEN SHOP-AND-SERVICE VENUES

NAME	NUMBER OF UNIQUE USERS	SUBCATEGORY
Crown Center	7,421	Mall
Zona Rosa	5,280	Mall
Trader Joe's	4,128	Grocery store
Apple Store Country Club Plaza	4,030	Electronics store
Cosentino's Market Downtown	3,703	Grocery store
H&M	3,361	Clothing store
Target	3,055	Department store
Barnes & Noble	2,901	Bookstore
MCI Car Rental Center	2,596	Rental car location
Marsh's Sun Fresh Market	2,544	Grocery store

TOP TEN FOOD VENUES

NAME	NUMBER OF UNIQUE USERS	SUBCATEGORY
McCoy's Public House	6,039	Brewery
Fiorella's Jack Stack Barbecue	5,809	BBQ joint
City Market	5,698	Farmers' market
Grinders	5,637	Pizza place
Cheesecake Factory	5,223	American restaurant
Beer Kitchen No. 1	4,866	Gastropub
Brio Tuscan Grille	4,774	Italian restaurant

Blanc Burgers + Bottles	4,767	Burger joint
Kona Grill	4,686	American restaurant
Jack Stack Barbecue Country Club Plaza	4,617	BBQ joint

The ten most popular Foursquare venues in Kansas City are found mainly in the areas near J. C. Nichols's Country Club Plaza. Most of these venues are local restaurant locations, such as McCoy's Public House, Fiorella's Jack Stack Barbecue (one of the better known barbecue establishments in the city), Grinders, Zona Rosa, the Cheesecake Factory, Beer Kitchen No. 1, Brio Tuscan Grille, and Blanc Burgers + Bottles. The Crown Center shopping mall holds the top spot for having the largest number of users check in, and the local City Market farmers' market holds the fourth spot (see table 7-1 for the top ten food and shop-and-service venues and their subcategories and number of unique user check-ins).

As table 7-2 shows, there is a significant correlation between the number of venues found on Foursquare and the number of people who identify as ethnically white or African American. While the areas identified as white have a significantly positive relationship with the number of businesses listed on Foursquare, the African American areas have a significantly negative relationship. The number of Foursquare venues is also significantly positively correlated with median family income, a relationship that is likely tied to ethnic characteristics.

Looking at the census tracts in Kansas City where the median age is either above forty-five or under twenty-five shows different trends in consumption (fig. 7-3). While some of the census tracts with older residents are located along the Kansas-Missouri state line, where the highest concentration of venues is located, there are several tracts where the median age is above forty-five and there are no Foursquare venues. Within the census tracts with older population there are forty-three Foursquare venues. Of the top ten most commonly visited venues only three

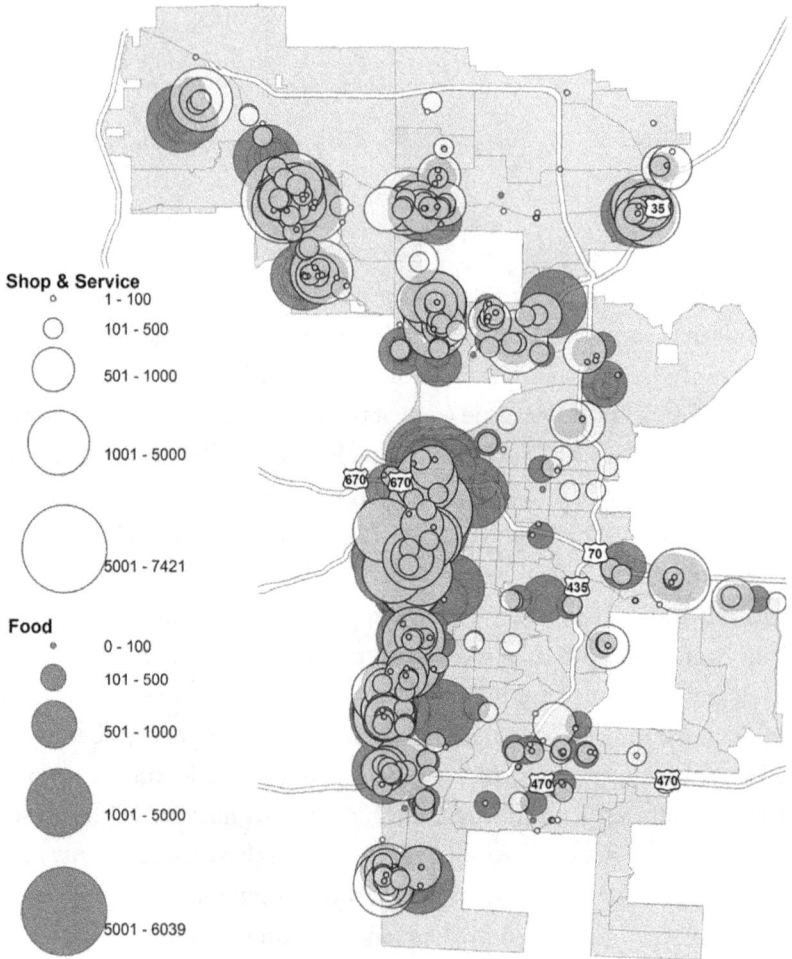

Legend:

Shop & Service
- ∘ 1 - 100
- ○ 101 - 500
- 501 - 1000
- 1001 - 5000
- 5001 - 7421

Food
- • 0 - 100
- 101 - 500
- 501 - 1000
- 1001 - 5000
- 5001 - 6039

FIG. 7-2. Foursquare food and shop-and-service venues in Kansas City, Missouri, by the number of users. Created by Emily Fekete.

TABLE 7-2. Correlations between Foursquare venues and selected demographics

	PROBABILITY VALUE	PEARSON CORRELATION
Total population	.242	.069
Median age	.297	.053
White	.019*	.205
African American	.005*	−.250
Hispanic	.136	−.109
Median family income	.013*	.219

*Significant at the .05 level.

are shop-and-service venues, with the remainder being restaurants. The venue with the most check-ins in the census tracts with older residents is a Quik Trip gas station and convenience store. Other shopping locations include a Price Chopper and Royal Liquors. The restaurants are mainly fast food and other chain restaurants and include Panera Bread, Applebee's Bar and Grill, McDonald's, and Taco Bell. The remaining three restaurants are Trezo Mare Restaurant and Lounge, Jasper's Italian Restaurant, and the Café in Briarcliff. This list of venues suggests that older populations may have a higher disposable income, so the tracts with older residents may be good locations for business owners to open restaurants.

Surprisingly, given the adoption of social media technology among younger groups, there are only six Foursquare venues listed in census tracts with a median age of below twenty-five. The only shopping option in these tracts is a Hy-Vee gas station, possibly because of a lack of clientele with substantial incomes in the area. The five restaurants represented are a Gates Bar-B-Q (popular local barbecue establishment with several locations throughout Kansas City), a Sonic Drive-In, a Dairy Queen, and two

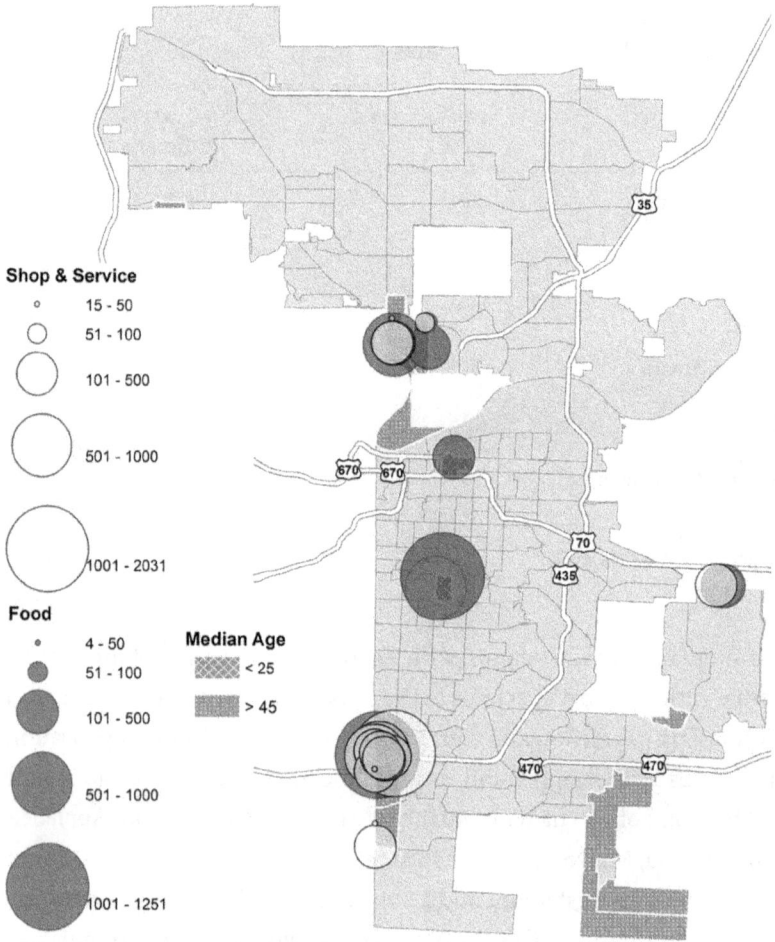

FIG. 7-3. Foursquare food and shop-and-service venues in Kansas City, Missouri, by the number of users in census tracts where the median age is either less than twenty-five or greater than forty-five. Created by Emily Fekete.

local places—Go Chicken Go, a car service station-turned-fried-chicken restaurant, and Pho Hoa, a Vietnamese restaurant. Based on the types of venues present, it may be that the census tracts with younger populations do not appeal to potential business owners as places to establish themselves or that young people cannot afford the rents in more developed census tracts.

There is a sharp contrast between areas of high income (census tracts where median family income is greater than $100,000) and areas of low income (census tracts where median family income is less than $20,000) in Kansas City, Missouri (fig. 7-4). Of the forty-eight total venues in high-income census tracts many are dispersed much like those found in older census tracts along the Missouri-Kansas state line, further supporting the point that areas with older residents also have residents with larger disposable incomes. Within the top ten most popular venues three are shopping areas, including a Target, a Best Buy, and Cosentino's Brookside Market—a high-end grocery store. The food venues that are present in the area do not necessarily reflect the higher earnings of the local population. Food venues included the Roasterie Café (a local coffee roaster and shop), Avenues Bistro, Julian, Steak 'n Shake, Panera Bread, Aixois, and Chipotle Mexican Grill. The dynamic of these retail locations in high-income areas is a product of the two distinct locations of high-income census tracts. The Target, Best Buy, Steak 'n Shake, Panera, and Chipotle are in a shopping plaza in northeast Kansas City, while the higher-end Roasterie Café, Julian, Avenues Bistro, Aixois, and Cosentino's are near the Country Club District on the Kansas-Missouri state line. The difference likely reflects the distinct types of high-income earners in each of these locations.

Low-income areas in Kansas City do not have many Foursquare venues, as only three are found in census tracts with a median family income below $20,000. The lack of venues could be due to the fact that lower-income families do not have smartphones and cannot use Foursquare. It could also be a reflection of a lack of retail trade outlets in areas with little affluence. Of the three venues in the area, one is the Vietnamese restaurant Pho Hoa found in the younger population census tracts. The other two are a Fast Stop gas station and a McDonald's.

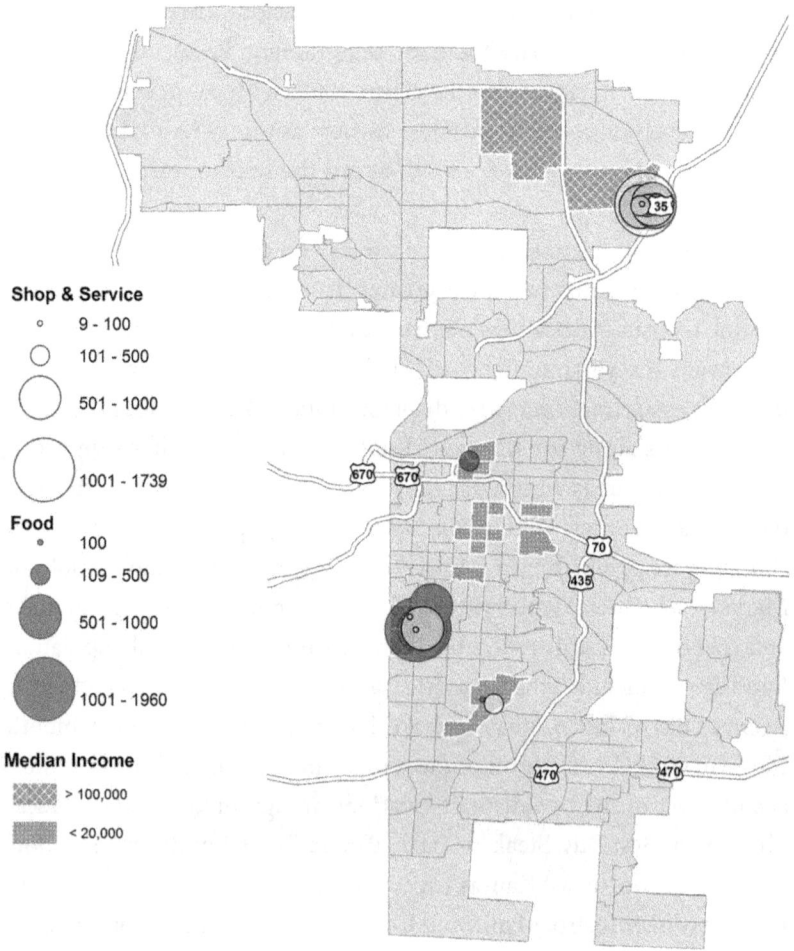

FIG. 7-4. Foursquare food and shop-and-service venues in Kansas City, Missouri, by number of users in census tracts where the median family income is either less than $20,000 or greater than $100,000 (unclosed lines mark edges of census tracts outside of Kansas City). Created by Emily Fekete.

Figure 7-5 is possibly the most reflective of the distinct ethnic divide that has long existed in Kansas City. While there were no census tracts that had a population that was 75 percent or more Hispanic, there are three where the majority population (greater than 50 percent) is Hispanic. There also are numerous census tracts in the inner city where the population is at least 75 percent African American, largely reflecting the historic pattern of settlement due to subsidized housing availability. The clearest example of the historical development of Kansas City is the location of census tracts where the white population makes up 75 percent or more of the total population in the area. These areas are clearly located along the state line. The white census tracts also contain most of the retail trade and food outlets found in Kansas City, with 584 Foursquare venues. The top ten most popular Foursquare venues in Kansas City as a whole are the same venues as the ten most popular Foursquare locations in predominantly white census tracts: Crown Center, McCoy's Public House, Fiorella's Jack Stack Barbecue, City Market, Grinders, Zona Rosa, the Cheesecake Factory, Beer Kitchen No. 1, Brio Tuscan Grille, and Blanc Burgers + Bottles. The development of the city spurred by J. C. Nichols is still apparent in the pattern of retail and ethnic segregation. This dispersion of venues found on Foursquare also reflects the economic divide that exists within the city, with more high-end establishments located in areas with a white and wealthy population.

Within the predominantly African American census tracts, there were only nine Foursquare venues. Of these nine the top two most frequently visited are the original locations of two Kansas City barbecue stalwarts: Arthur Bryant's Barbeque and Gates Bar-B-Q. It is likely not locals who are checking in to these two locations on Foursquare but rather barbecue enthusiasts and tourists who are artificially driving up the number of users who have checked in. There are three other restaurants in the African American census tracts: Niecie's Restaurant (a local diner), a Sonic Drive-In, and a Taco Bell. There is one option for shopping (a CVS pharmacy), two banks (Bank of America and Mazuma Credit Union), and the Southeast Community Center. The lack of Foursquare venues in these areas could be either a product of a lack of engagement with Foursquare by African Americans or the fact that these areas do not contain any venues where one may check in. On the ground analysis of these areas is incredibly important in order to determine the extent of assumed food deserts and lack of access to other necessities in these census tracts.

Census tracts with a majority Hispanic population have more Foursquare venues than do African American census tracts, but these venues might also exist because of people coming in to the tracts for specific types of food or for work. Although there are nineteen Foursquare venues listed for these areas, only three of them are shop-and-service venues (with one grocery store, a Price Chopper) with the rest being restaurants or bars, also possibly creating an urban food desert for many residents. The popular venues in this area reflect the history of these census tracts as an old industrial area of the city. Both Boulevard Brewing Company and the Roasterie Coffee Plant have their operations in these census tracts and are among the ten most visited venues. The Boulevard Tasting Room also made the top ten venue list. There are two local farm-to-table restaurants, Westside Local and Blue Bird Bistro, and a creperie, Chez Elle, in these census tracts, and the remaining four restaurants are local Mexican places: Ponak's Mexican Kitchen & Bar, El Patrón, Margarita's, and Los Tules. It would be worth investigating gentrification in these census tracts, as they are located in old industrial areas and seem to have a mix of hipster and Hispanic culture.

Data and Demographics: Connections or Coincidence?

The development of retail trade establishments in Kansas City is largely a reflection of the historical trends put into place by people such as J. C. Nichols and projects such as his Country Club Plaza. Tied to the urban development that occurred during the twentieth century is a history of ethnic segregation that is still evident through census tract data. The historical development of the city can clearly still be seen through a combination of census tract data and Foursquare check-in data for food venues and shop-and-service businesses. There is a clear core area of Kansas City that houses a majority African American community, while the area around the state line is largely white. Although there is a growing Hispanic presence,

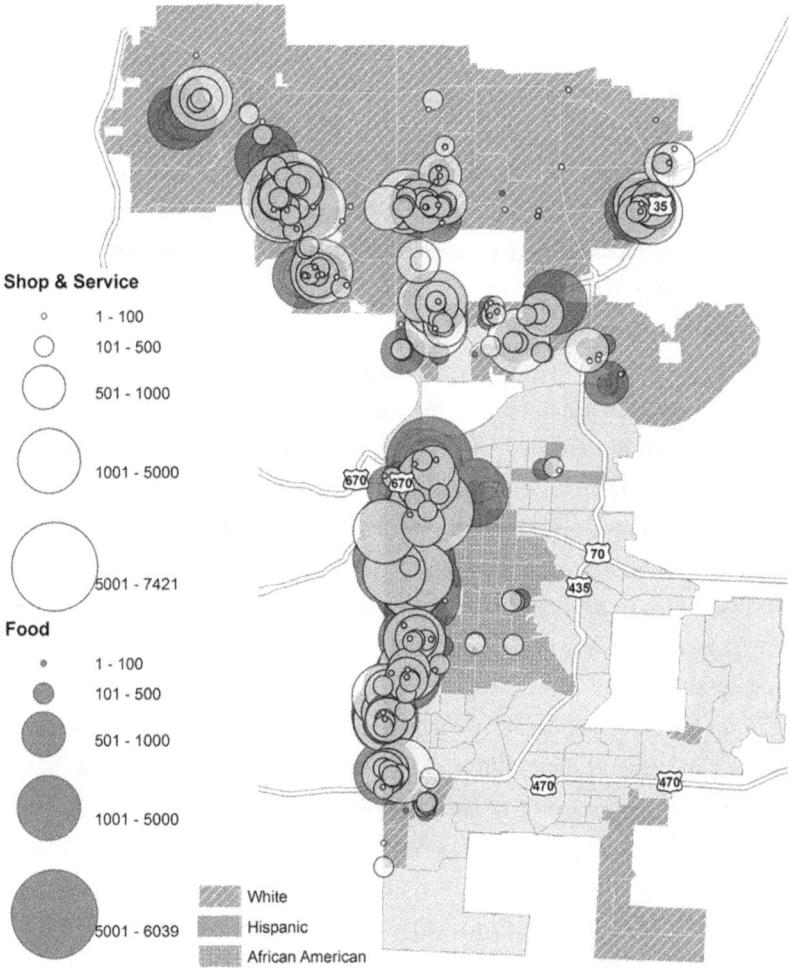

FIG. 7-5. Foursquare food and shop-and-service venues in Kansas City, Missouri, by the number of users in census tracts where the majority race is either white (greater than 75 percent), African American (greater than 75 percent), or Hispanic (greater than 50 percent). Note: tracts with no majority are not highlighted. Created by Emily Fekete.

it has not yet manifested itself on the ground in a significant way, though there is evidence of this minority population through some of the dining establishments found on Foursquare. The ethnic divisions of the city mirror the availability of goods and services found on Foursquare. African American census tracts do not have the numbers or diversity of products that other census tracts do. Retail trade is skewed toward the white and the wealthy. While the distortion of available Foursquare venues could be a reflection of the population using Foursquare, it could also point to the lack of businesses that exist in African American communities.

In looking at the patterns of Foursquare use outlined above, it can be deduced that the application appeals mostly to a white population. The probability of Foursquare users being white is evident in both the location of venues found along the Kansas-Missouri state line as well as the lack of businesses on Foursquare in African American neighborhoods. There are serious real-world implications for the lack of representation of businesses in African American census tracts on Foursquare.[34] With the Explore feature on the application, Foursquare users will see only businesses listed online. People using Explore to find a retail location will not have the opportunity to see businesses in predominantly African American areas of a city. Similar to the divide found on Google Maps by Zook and Graham or in New Orleans by Crutcher and Zook, Foursquare presents a specific version of a city to its users, one that largely overlooks the ethnic composition of businesses.[35] This is not to say that African Americans should or need to start using Foursquare; however, it does establish the political position that Foursquare presents to its users. Overlooking an entire ethnic category of the population does not create an ideal environment for people to "discover new places and explore what is nearby," as the service claims.[36] It offers instead a biased reproduction of retail locations based on a specific sector of the U.S. population who visit these establishments. The places that are viewed on Foursquare as the most popular places to eat or shop in a given location may in fact not be as popular when looking at the ethnic character of a city.[37]

Tying social media data strictly to demographic data is challenging. Looking at social media data through a lens of demographic information has the potential to illuminate on-the-ground problems, but it likely only further points to a topic often overlooked by those studying big data: the digital divide. While there is a possibility that some of the census tracts in Kansas City (notably the African American tracts) are underrepresented in terms of access to retail establishments, what is more likely is that the data on Foursquare are being created by a nonrepresentative group of users. Despite the ongoing myth of an all-inclusive Internet, digital divides continue to persist within the United States.[38] Applications such as Foursquare that have filters built into them (like Explore) are largely a reflection of the users who have developed the application, whether it is open source or not. Without knowing it, people may be basing their consumption decisions on a computer code that does not have the ability to recognize what has been omitted from its database. As people increasingly rely on technology to direct their choices of where to go for obtaining goods and services, those businesses left out because of the technological divide will suffer.

What big data does illuminate are areas that may require a more in-depth analysis that cannot be gleaned from digital data sources alone. Big data can provide a social context in which to understand large-scale unevenness in access to sites of consumption. Often this access is a product of greater societal issues that could stem from historical development or the digital divide. However, in attempting to address these concerns, researchers must not only rely on what they uncover from an initial big data survey but use a variety of methodological approaches in order to address the issues present. Researchers should be wary of relying too heavily on user-generated social media data without recognizing the inherent biases built into these sources.

Notes

1. Barnes, "Retheorizing Economic Geography."
2. Hartwick, "Geographies of Consumption"; Valentine, "Corporeal Geography of Consumption"; Miller, "Could the Internet Defetishise?"; Tokatli, "Old Firms, New Tricks."
3. Dalton and Thatcher, "Inflated Granularity"; Singleton and Spielman, "Past, Present, and Future."

4. Fekete, "Consumption and the Urban Hierarchy."

5. Foursquare, "About."

6. Foursquare, "About."

7. Popper and Hamburger, "Meet Swarm."

8. Hern, "Why Foursquare."

9. Popper and Hamburger, "Meet Swarm"; Gordon and de Souza e Silva, "Urban Dynamics of Net Localities."

10. Zook and Graham, "Mapping DigiPlace"; Stephens, "Gender and the Geoweb."

11. Kitchin and Dodge, *Code/Space*.

12. Stephens, "Gender and the Geoweb."

13. Stephens, "Gender and the Geoweb."

14. Stephens, "Gender and the Geoweb," 982.

15. Rose, *Feminism and Geography*.

16. Crutcher and Zook, "Placemarks and Waterlines."

17. Crutcher and Zook, "Placemarks and Waterlines," 523.

18. Stephens, "Gender and the Geoweb."

19. Pew Research Center, "Mobile Fact Sheet"; M. Anderson, "More Americans Using Smartphones."

20. Smith, "46% of American Adults"; Fox and Rainie, *Internet at 25*.

21. Foursquare, "About."

22. Lindqvist et al., "I'm the Mayor of My House."

23. U.S. Census Bureau, *American Community Survey*.

24. Massey, Gross, and Shibuya, "Migration, Segregation."

25. U.S. Census Bureau, *American Community Survey*.

26. Historic Kansas City, "Our City."

27. Historic Kansas City, "Our City."

28. Shortridge, *Kansas City*.

29. Gotham, "Urban Space."

30. Gotham, "Urban Space."

31. Shortridge, *Kansas City*.

32. Gotham, "Urban Space."

33. Kansas City Convention and Visitors Association, "Kansas City."

34. Fekete, "Race and (Online)."

35. Zook and Graham, "Mapping DigiPlace"; Zook and Graham, "Creative Reconstruction"; Crutcher and Zook, "Placemarks and Waterlines."

36. Foursquare, "About."

37. Fekete, "Race and (Online)."

38. Morozov, *To Save Everything*; Malecki and Moriset, *Digital Economy*.

PART 4

Urban Big Data

Urban-Centric and Uneven

8 Big City, Big Data

Four Vignettes

Jessa Lingel

As someone who has lived her entire adult life in cities and as a researcher who studies spatial informatics, I have often felt a combination of guilt and concern at having contributed to what Halberstam has called metronormativity—the overemphasis on urban contexts at the expense of other places, such as suburbs and rural areas.[1] As a global trend, urbanization has important social, economic, and environmental consequences that demand scholarly attention. And yet when cultural geographers take cities as their object of study, it becomes very easy to conflate urban experience with human experience, further entrenching stereotypes that people outside city limits are also outside the purview of scholarly work. So it is with a somewhat rueful acceptance that I admit the utility of thinking about cities in order to understand big data in the context of everyday life.

It would be fair to say that big data skews toward two loci of social concentration: major social media sites and major cities. I will not expand here on the hyperconcentration of scholarly attention on just a few online platforms.[2] This is partly because the focus of this edited volume is on big data and space and partly for reasons of protecting privacy; I sometimes think that academics (and corporations) should continue to ignore smaller social media platforms precisely because they are increasingly the only sites of online interaction that are afforded any measure of privacy simply because they are ignored. I will instead focus on lived experiences of big data as a way of pointing toward ethical contours of city life and statistical paradigms of measurement.

It is almost impossible to think of big data as material, as accessible in a physical way, but somewhat easier to think in terms of its materiality.[3] By this I mean the ways that big data projects alter lived experiences of the city. To ground (or materialize) this convergence of the urban and the algorithmic, I offer four vignettes of encountering big data in everyday urban life.[4] These vignettes emphasize themes of measurement, infrastructure, legibility, and play. I conclude with some thoughts on how these experiences have shaped my own beliefs of the ethics of big data and urban informatics. By connecting big data to life in a big city, I do not mean to imply that non-urban areas are immune to the reach of big data. And yet being part of a crowd has long been hailed as the primary and emblematic means of offering city dwellers anonymity.[5] In this short chapter I am concerned precisely with experiences of realizing how this anonymity is being methodically and invisibly dismantled to the advantage of corporations and at the expense of privacy, community, and play.

In the late summer of 2013 I am attending an urban planning workshop in New York City. Bringing together NGOs, policy makers, social scientists, and industry people, the workshop is intended to advise a partnership between a development project (whose goal is to build a retail center on the west side of Manhattan) and a university department (whose goal is to become a global leader in big data initiatives at the convergence of academia and industry). As a qualitative researcher and a woman, I find my field and my gender to be underrepresented in the room. As an activist and Marxist feminist I feel all but illegible, but then again so are some of the statements made in the course of explaining the different data-gathering projects being conceptualized. Men in suits talk about the things they will measure with sensors: "We can monitor foot traffic and energy use! Trash! Alcohol consumption! Smoking! Sleep!" I realize I am watching big data happen. The development initiative includes both residential and commercial property; the goal of all this measurement is allegedly to gauge "quality of life." I voice concerns about privacy.[6] I also bring up alternate methods of inquiry and mention my doubts that any of these things will in fact add up to the stated goal of understanding whether people are happy. I am told, "Our goal is not to have to talk to people face to face" and then, "You cannot improve what you cannot measure."

I'm on a G train in Brooklyn, and the year is 2011. On one side of the train I see advertisements for Google's Good to Know initiative, which informs users on issues of privacy, security, and netiquette.[7] For example, one ad advised against the perils of lazy password selection: "Choosing the same password for each of your online accounts is like using the same key to lock your home, car and office—if a criminal gains access to one, all of them are compromised. So don't use the same password for an online newsletter as you do for your email or bank account. It may be less convenient, but picking multiple passwords keeps you safer." I turn my head and on the other side of the train, I see promotions from the Metropolitan Transit Authority, which runs the public transit and port authorities in metropolitan New York. The ads simultaneously document and advertise progress on various service improvements, assuring us that the Second Avenue subway line is finally coming, that the Fulton Street station will eventually open, that security cameras are being installed on city buses for our "safety." It occurs to me that these ads are doing the same thing, which is to say they are presenting institutions as infrastructure. Part of this collapse comes through generalization—the benevolent distribution of information on safety and progress. There is a shared emphasis in these messages on proactive responsiveness ("Take note of our efforts to meet your needs!") that also erases competing platforms and services. Google is not the Internet; it is a company that produces tools (email, search engine, cloud storage) that facilitate online tasks. The MTA is not public transportation; it is a public benefit corporation that relies on and maintains infrastructure that long predates the MTA's existence as an institution.

Typically I am very wary of using spatial metaphors to describe web-based technologies.[8] The connection is useful here, however, in that both Google and MTA are invested in presenting themselves as integral to routine travel, whether online or off. This conflation is all the more disconcerting when we think about flows of information and data and bodies that are managed and facilitated by institutions acting as if they are infrastructure. When pundits point to the popularity of tools like Facebook and EZ Pass as indications that people must not care about privacy (since they give so much personal information to these companies), they fail (or refuse) to acknowledge that, for many people, not using these technologies doesn't feel like an option.[9] And part of the reason that opting out doesn't feel like a choice is that entities like Google and the MTA present themselves not as one product among many but as ingrained, inevitable infrastructure. As I trundle along on the G train in Brooklyn these ads want to reassure passengers that these companies are continuing to work for our benefit, safety, and convenience but less, perhaps, as consumers than as the beneficiaries of a public good.

It's the summer of 2004 and I have just moved to New York. On my first night in the city that will be my home for the next decade, I go with my new roommate to a bar called the Metropolitan. It is only when I read a Sleater Kinney lyric scrawled on a bathroom stall door that I realize two things: (1) I'm in a lesbian bar and (2) I can make a home out of this city. Writing this now, I realize that this is my first memory of reading New York. By "reading" I'm referring to the interpretive work of spatial sensemaking. In her book *Metropolitan Lovers* Julie Abraham unpacks the (specifically queer) history of this kind of reading, writing that "anxieties about the legibility of buildings and streetscapes, about being able to interpret the cities and versions of cities that we see . . . are also an extension of the anxiety about legibility [of otherness]. . . . It is a [modern] anxiety about the 'facades' of individuals rather than of buildings, about the artifice of urban life undermining the legibility of persons in the city. Postmodern anxieties about the reading of urban structures echo the anxieties of the denizens of great cities about the readings of persons."[10]

It is worth returning to these anxieties of bodies, buildings, and desires relative to concerns about big data.[11] With locative media and geo-located photo tags, moving through everyday urban life increasingly produces (both actively and passively) data that accumulate powerful sets of information of interest to corporate as well as government actors. In other words our movements make the city legible to entities that can then profit from our statistical contributions. And yet legibility can function both top down and bottom up. Queer women make themselves and their spaces legible to each other and create a sense of home, even as we simultaneously become legible to those who would do us harm. Legibility has been a rallying cry of historically marginalized groups for a long time; I am cognizant of the need to honor those efforts, even as they come with new concerns (of privacy as much as censorship) when relationships to data are centered on prediction and profit first and on interpersonal contact and community last, if at all.[12]

Just two months shy of my nine-year anniversary in the city, I leave New York for Boston. The week of my departure the city installs CitiBikes, its bike-share program. I feel an unexpected sense of relief that I do not have to see (or bike through) the city this way. One of my favorite things about urban biking is the feeling that I am no longer bound by subway lines and bus routes, that I have far more agency in how I navigate streets and neighborhoods, basically that the city can feel like play. CitiBikes, which incidentally represent quite neatly the conflation of institution and infrastructure, are bound temporally (to thirty-minute rides) and spatially (no bikes for the vast majority of Brooklyn and Queens, or the entirety of the Bronx, and Staten Island). They are also binding cyclists in data sets that become public but not transparent.[13] It's one thing to be able to download an Excel spreadsheet loaded with CitiBike statistics and quite another to offer APIs and visualization tools required to make sense of and play with these statistics. One's presence in a data set moreover is never neutral and never without consequences (however small) for other points of data. Big data is tied to cities in that density is necessary for analysis; outliers and patterns become visible only with an adequate baseline of normative datapoints. In this way legibility of data is not only (and in fact not at all) individual; it is collective. It is of course a choice to use CitiBike, just like it's a choice to ride the subway. What CitiBike illustrates so clearly is how movements through city space articulate relationships of privilege, privacy, and monitoring. These layers of legibility—between institution and infrastructure, between legibility and enclosure, between individual and collective—are what make it increasingly difficult to retain a sense of play in one's daily movements through city space.

Despite many, many experiences of this not being the case, I continue to subscribe to a vision of the city as a place where play is possible, even fostered. In her work on privacy regulations Julie Cohen has argued that privacy law should fundamentally respect human creativity, facilitating socio-technical forms of play.[14] In thinking both about policy and research initiatives for big data and city space, I believe Cohen's key claims should drive decisions on how we gather data about everyday urban life. These decisions have consequences—I worry about the chilling effects of big data in the context of mobility, privacy, and play. I think of the intrusiveness of censors that residents and workers can't see or contest. Ideals of play don't hold up against the fatigue of encountering technologies that feel inevitable and entrenched, with profoundly uneven distributions of access to information.

Virginia Eubanks has written persuasively and powerfully on the ways that economically disempowered and disenfranchised groups have long experienced daily acts of monitoring and surveillance that increasingly define contemporary urban life for the poor, such as the tracking of food consumption, the number of people in a household, and records of medical visits and treatments.[15] It seems increasingly obvious that these same infrastructures of tracking and monitoring will continue to be ingrained into technologies of everyday life across socioeconomic strata. To some extent, this has been the case for a very long time, as evidenced by credit card reports and grocery store customer loyalty programs.[16] In the past corporate entities seemed reluctant to acknowledge this statistical work as surveillance, fearing customer backlash related to privacy concerns.[17] The technological capacities for data gathering and algorithmic calculation are typically cited as catalyzing forces for the emergence and power of the big data paradigm, and yet there is another, discursive shift that seems to have taken place in which tools of surveillance that used to be viewed as creepy are now lauded as revelatory.[18] In other words, rather than an intrusion on privacy, data analytics are positioned as a gift of insight.

I am wary of seeming overly negative (some would say paranoid) about uses for big data in the context of city life. Surely it's possible for tools of big data to be tools of curiosity and equality rather than disempowerment.

I therefore conclude by suggesting means of intervention in the cause of ethical urban informatics. I first propose considering alternate practices of legibility—ways of knowing and reading the city that aren't rooted in the endless proliferation of unseen, un-asked-for tracking. For example, Jack Gieseking's combination of social network analysis and meticulous archive work on queer advocacy and community groups in New York provides a fascinating look at how queer organizations emerged over time, with implications for deeper understandings of the convergence of spatial and political policy.[19] Gieseking's approach is very much tied both to data analytics and to goals of legibility, bringing to light social and institutional connections that would otherwise remain hidden. Admittedly this methodology is an archival one; for research that has a more immediate time frame, legibility could be considered a process of interpreting ephemera (the traces left behind in the course of everyday life) as ephemeral, meaning something to be interpreted and analyzed but not necessarily reinscribed in an archive that is inaccessible or inscrutable to the people whose data it contains. Part of ethical big data initiatives requires more parity in agency, first by allowing people to opt in and out of participation and second by allowing data to disappear. When the CitiBike program offers its clients data sets comprising their own data points, the gesture rings hollow partly because users have limited agency in how their data live within that database and partly because the data sets themselves are overwhelming to someone without training in statistical methods.[20]

In a second set of demands for ethical approaches to big data, I want to advocate for committed attention to social justice as fundamental to how processes of statistics, measurement, and algorithms are experienced in daily city life. Partly this commitment should take place through an insistence on ethics training for academics working in computational social science (and related subfields) who may not realize the extent to which their work on big data is often work that involves human subjects. Internal review boards (IRBs) tasked with evaluating the legal and ethical implications of research with human subjects must combat tendencies to lag in their familiarity with contemporary research tools and trends, including the APIs and algorithmic analysis used in big data research. On a policy level the work of IRBs cannot be limited to academics; as noted earlier there have always been compelling motivations for corporate and government actors to conduct consumer and civic analysis, but the scale and lack of transparency around big data collection present new and troubling issues in protecting privacy, agency, and creativity. Policy makers ignore the ethical implications of these issues at their and their constituents' peril. As researchers and technologists we must commit ourselves to making these processes of data gathering more visible, to articulating consent clearly and honestly rather than wearily surrendering to the highest common denominator of legal language, to supporting complex and even contradictory desires for mobility as well as stasis, and to imagining different political arrangements that secure the right to work as well as play.

Notes

1. Halberstam, *Queer Time and Place*. See also Gray, *Out in the Country*.
2. To offer a brief, illustrative anecdote, I was once told by a former Twitter intern that the company estimated that at that time (summer of 2012) *every* public tweet had been included in a data set as a point of analysis. That sites like Craigslist, LinkedIn, and eBay go largely unexamined while Facebook, Wikipedia, and Twitter often get multiple panels within a single conference demonstrates the myopia of Internet studies researchers.
3. Dourish and Mazmanian, "Media as Material."
4. These narratives are all based on my nine-year residence in New York, which is its own form of myopic metronormativity in that New York is often held up as the über-city that can stand in for all other cities.
5. Benjamin, *Arcades Project*; Simmel, "Metropolis and Mental Life."
6. Upon questioning, I was told that consent was determined by the lease, meaning that residents would be informed of sensing devices, but employees wouldn't necessarily be informed of the rental agreements signed by their employers.
7. Google, "Good to Know."
8. See Chun, *Control and Freedom*, particularly chap. 1; Donath, *Social Machine*, particularly chap. 3; Stefik, *Internet Dreams*; and M. Graham, "Geography/Internet."
9. See Nissenbaum, *Privacy in Context*, 149–50.
10. Abraham, *Metropolitan Lovers*, 36.
11. See Crawford, "Anxieties of Big Data."
12. On prediction and profit, see Andrejevic, *Infoglut*.

13. CitiBike, "System Data."
14. J. Cohen, *Configuring the Networked Self*.
15. Eubanks, *Digital Dead End*. For a historical perspective, see Abraham, *Metropolitan Lovers*, 205.
16. Pasquale, *Black Box Society*, details credit card reporting.
17. Duhigg, "What Does Your Credit-Card Company Know?"
18. For example, General Electric, "Datalandia Trailer—GE," a movie trailer spoof on Big Datalandia: "In a world plagued by menacing aliens, sexy vampires, and cornball romantic comedies, one small town must rely on a new kind of hero: Brilliant Machines. Welcome to Datalandia, the little town saved by Big Data." That GE chose to locate its narrative in a "little town" rather than a city perhaps asks for a framing in the "Internet of Things" rather than Big Data discourse, but that's beyond the scope of this paper.
19. Gieseking, *Queer New York*.
20. There are, however, minor acts of pranking that in some way form a kind of counterconduct, such as a Nike+ user who effectively drew a penis with his running route uploaded to a map of San Francisco, as documented in Parkinson, "Runner Uses Nike+ App."

9   Framing Digital Exclusion in Technologically Mediated Urban Spaces

Matthew Kelley

Scholarly work on digital exclusion and inequality (often referred to as the "digital divide") reached a crescendo during the late 1990s and early 2000s. If we look back it is no surprise that, given the pace of innovation and the emergent nature of consumer-oriented digital technologies, there would be uneven patterns of adoption and usage among the general population. Urban areas in particular were looked to by early digital divide scholars as sites within which the exclusivity of digital technologies was most geographically pronounced. But as patterns of usage stabilized and as the general population became more comfortable with the integration of digital technology and everyday life, critical scholarly and policy-oriented attention has shifted away from the divide as it was initially conceived. In a manner of thinking, while once the divide had seemed impossible to bridge, it became over time little more than a crack in the socioeconomic fabric of urban space. By the mid-2000s Internet access and personal computers were nearly as ubiquitous in U.S. cities as telephones and televisions. Access gaps that remained in disadvantaged neighborhoods were partially filled through public investment in technology centers, public libraries, after-school programs, and the like. Although there remained significant efforts to train socioeconomically disadvantaged residents to use digital technologies, even these efforts became less urgent as a generational shift occurred. Children of the 1990s became young adults with computers and Internet connections in the background of their lives. Toward the end of the first decade of the twenty-first century, however, the ongoing evolution of digital technology began to open an interesting new series of fissures in the socio-technical landscape.

Computers became smaller, phones became mobile, and Internet connections became pervasive. Whereas the digital divide of the late 1990s was defined almost entirely by the question of access to a computer with a modem, exclusion and inequality in the socio-technical landscape of the 2010s is more nuanced—tied in large part to the increasingly passive mediation of everyday life by *smart* technologies. The field is complicated by the ever-growing stable of consumer-oriented gadgets that are embedded with microprocessors, locative capacities, and interactive functionalities. Social exclusion as a result of the prohibitive financial burdens associated with ownership of digital technology is no longer foreground. Instead we turn to issues such as class-based differentiations in mobile technology usage and the impact that geosocial information can have on place. Much contemporary work focuses primarily on the data and information that are produced and consumed by mobile digital technologies while connected to pervasive Internet services. Digital divide research, in short, is in the midst of a focal adjustment, and there remains an open question as to how we might best refocus efforts to understand and engage with the unexpected consequences of new technologies in urban space. In this chapter I situate current and potential critical divide-esque scholarship alongside the work that has been ongoing since the late 1990s. My aim here is to offer this overview of past and present divide-related scholarship while also encouraging new work on the less visible (and arguably more complicated) contemporary uneven socio-technical landscapes. I begin by providing a brief background on scholarship that has focused on digital exclusion and inequality and then move into a discussion of the new range of technologies that began to pervade everyday life toward the end of the first decade of the 2000s. In the final section of the chapter I reengage the problem of digital exclusion and inequality from within the contemporary socio-technical framework— thinking through not only how we might reimagine what a digital exclusion is but also how the consequences of more subtle and pervasive digital exclusion might be manifest in digitally engaged urban spaces.

Digital Exclusion and the Divide, Late 1990s–Early 2000s

In the mid-1990s the U.S. National Telecommunications and Information Administration (NTIA) set out to study the changing patterns of digital information access across the United States.[1] Historically the NTIA had been charged with working toward achieving the U.S. telecommunications policy goal of universal and affordable telephone service. But in a 1995 report, cleverly titled *Falling through the Net: A Survey of the "Have Nots" in Rural and Urban America*, the NTIA suggested that its mission had begun to change: "While a standard telephone line can be an individual's pathway to the riches of the Information Age, a personal computer and modem are rapidly becoming the keys to the vault. The robust growth recently experienced in Internet usage illustrates this promise as new and individual subscribers gravitate to on-line services. This suggests a need to go beyond the traditional focus on telephone penetration as the barometer of this nation's progress toward universal service."[2]

Responding to the relatively rapid diffusion of computers and Internet connectivity in the 1990s, the NTIA expanded its scope to include digital information dispersed through Internet-based communications. So, just prior to releasing the 1995 report, the NTIA contracted with the U.S. Census Bureau to begin collecting demographic data related to computer and modem ownership as a way for the agency to gain "new insights about the 'information disadvantaged' in America's central cities."[3] Evident here should be the emphasis on *access* that permeated early divide-related U.S. policy documents. The revised NTIA goal of universal and affordable access to the riches of the information age via computers with Internet access was baked into federal and state policy by the late 1990s. As I briefly detail below, the flurry of activity following the work of the NTIA (coupled with market-driven decreases in the costs associated with Internet access) was fairly successful in the provision of affordable access to *most* urban residents by the middle of the decade after 2000. What followed, however, was a new manifestation of digital exclusion that was characterized not by uneven access to digital technologies and information but by an unevenness in technical literacy that mirrored the socioeconomic schism of the access divide. The same social groups most disadvantaged by the access divide also had the greatest skills deficit once they gained access to the technologies. There were in other words two incarnations of the digital divide between the mid-1990s and about 2010—the first-order divide being characterized by access and the second order, by literacy.

The first-order digital divide was both conceived and engaged as a binary social problem of *haves* and *have nots*. Individuals who did not have the means to own a computer and connect to the Internet were excluded from the social and economic possibilities engendered by digital information. By placing a computer and Internet connection into the homes of the most socioeconomically disadvantaged, policy makers expected incredible outcomes—from ending chronic unemployment to leading whole communities out of the cycle of poverty. As Kruger noted, "it has become one of the orthodoxies of the late 1990s that the information and communication technology (ICT) revolution potentially offers one way of tackling almost every social ill."[4] By the year 2000 serious efforts were under way at all levels to guarantee access to the "information have nots" of urban areas. Community technology centers, after-school computer programs, and the expansion of digital technologies within public library systems were among state-sponsored efforts to expose the have nots to the brave new world of digital information.[5] The U.S. Federal Communications Commission (FCC) released a report in 2000 that detailed not only the disparity of access to broadband (defined at that time as in excess of 200 Kbps in at least one direction—upstream or downstream) in the United States but also several policy initiatives that could be leveraged to improve access.[6] Among the initiatives the report identified was municipal broadband in particular, which would be a (potentially) cost-effective and well-regulated source for affordable and universal Internet access. Alongside initiatives to provide access to computer hardware, municipal broadband efforts in the early 2000s were a key element in the vast state-sponsored "bridging the digital divide" agenda. The takeaway here is not, however, in the details of efforts to bridge the access divide. It is rather that once widespread digital exclusion had been linked to access in the mid-1990s, vast resources and immeasurable efforts were expended to eliminate the inequality that was manifest by the divide, leading, though not causally, to the second-order problem of digital exclusion as a result of uneven literacy.

Scholars have argued that low levels of digital literacy, as a form of social exclusion, can adversely impact residents' civic participation and sense of place.[7] Mesch and Talmud offered evidence of the positive effect that digital literacy and participation can have on real-world communities by drawing on data from longitudinal research in two Israeli communities.[8] In their work they found that digital literacy enabled people to participate in online activities, which increased both the level of civic engagement among residents as well as residents' sense of connectedness to their neighborhoods. On the flip side, in communities with low levels of literacy residents are excluded from digitally enhanced civic participation and place making. Unfortunately, unlike access to digital technology, uneven digital literacy was a more complex problem for policy makers to confront. Lindsay argued, for instance, that the predominantly low levels of digital literacy in disadvantaged urban neighborhoods were only exacerbated by the absence of technical sociocultural support mechanisms at the local level.[9] Cultivating such mechanisms was a neither quick nor easy task, and as is most often the case with community development efforts in socioeconomically disadvantaged places, the strategies and techniques that are employed in a particular neighborhood must reflect the uniqueness of that neighborhood's social, cultural, and economic characteristics.[10] Because low levels of digital literacy tended to adhere to the same spatial patterns as other socioeconomic challenges, such as high poverty and unemployment rates and low high school graduation rates, to fully engage the problem place makers—community development professionals, activists, urban officials, and the like—could not rely on macro-scale practices.[11] The process was necessarily a slow one because overcoming the literacy divide was as much a matter of teaching technical skills as it was a problem of infusing local cultures with new perceptions of digital technologies.

Efforts to overcome the literacy divide almost mirrored efforts to provide access in the late 1990s. But although strategic investments such as community technology centers, after-school programs, and adult education efforts were implemented in disadvantaged communities, digital literacy is largely characterized by tacit skills that are acquired through sustained and informal exposure to computers and the Internet. As the access divide was gradually overcome, a generation of youth received an introduction to digital technologies. As they became adults in the late 2000s and early 2010s, tacit digital literacy was increasingly apparent. As I suggest in the following section, though digital literacy remains uneven along class-based lines, more attention is being given to *how* individuals participate in the digital world than whether or not they are *able* to participate. Hargittai and Hinnant's investigation of "young adult" Internet users focused, for instance, on the differences in activities among a group of users who had access to digital technologies but who had varying levels of digital literacy.[12] Meyen et al. went so far as to offer a typology of Internet users—speaking not directly to varying levels of digital literacy but instead recognizing the reality of its unevenness.[13] Finally, it is notable that more recent work on digital inequality (from access to literacy) assumes a baseline level of literacy among most citizens in the developed world—basic digital skills, it is assumed, have become tacit through increased exposure to technologies since the mid-2000s.[14]

Emergent Technologies and Urban Space

Alongside the diffusion of pervasive Internet connectivity and the cascade of smart, mobile, and location-aware handheld digital devices, a good deal of scholarly attention has been oriented toward understanding how new domains of digital reality can integrate with material society and what consequences of this integration can be expected in terms of everyday urban life.[15] Following the theme of digital exclusion established in the previous section, there are a host of new challenges associated with the increasingly unencumbered diffusion of digital technologies across urban space. Just as the second-order divide was exponentially more difficult to overcome through the application of conventional urban policies than the access divide, there are arguably even fewer clear pathways to take when approaching new manifestations of digital exclusion and inequality in urban space. As I discuss throughout the remainder of this chapter, the complexity of the consequences of digital exclusion and inequality in urban space is a function of the almost limitless range of affordable computing devices that are both pervasively connected to the Internet and location aware. With greatly improved levels of access to digital technology residents' tacit literacy in urban space has improved dramatically, and as a result the topics or problems of digital exclusion and inequality have become less apparent. Inasmuch as the unevenness of digital literacy began to wane in the early 2010s, exclusion simultaneously became more a matter of *how* rather than *if* digital technologies mediate the experience of social and material reality. Unlike our approach to the problem of access to technology or digital literacy, when investigating the potential for a next-order divide we must begin by investigating the new modes that digital exclusion is likely to inhabit. For instance, do new modes of digital exclusion tend to adhere to the same social and geographic patterns as previous layers of digital exclusion? More significantly, how are the consequences of these modes of digital exclusion manifest? As I emphasize again toward the end of this chapter, our approach to any new manifestation of the digital divide will likely need to be much more creative than in the past, because coarse policies aimed at leveling the digital playing field will have little effect when the problem of digital exclusion has become so nuanced as to have fallen out of the popular consciousness.

Before engaging in a discussion of new modes of digital exclusion, I would like to take a brief step back in order to first establish the remarkable transformation that the digital landscape has experienced since 2010. I argue below that three characteristics of this landscape (user interactivity, pervasive connectivity, and location awareness) have been particularly notable as we dive more deeply into an investigation of the social and socio-spatial impact of new technology. Location-based, pervasively Internet-connected digital tools have gradually engendered new socio-technical possibilities as users are granted new forms of participation in the production of publicly accessible digital (and largely geospatial) information. Hudson-Smith et al. have referred to this phenomenon as "mapping for the masses" as for the first time in history control of geospatial information has been granted to anyone (or any public/private/not-for-profit entity) with a moderate digital skill set.[16] Scholars have since taken to referring to the agglomeration of locative (geospatial) tools and information on the Internet as the geoweb, and much contemporary socio-technical and geospatial research is framed using existing geoweb scholarship.[17] This research is particularly significant to emphasize here because it engages with the essence of the contemporary digital landscape—mobile, locative, and pervasively Internet-connected devices—which is the ability for an unobtrusive piece of technology to mediate and facilitate everyday life by maintaining context awareness and augmenting the user's spatial intelligence through subtle signals. Thrift and French were among the first to observe the oncoming geoweb when they predicted an emergent class of devices that would "be 'practice-aware,' responding to and aware of the context in which they are used . . . this means we are moving away from machines that simply respond to machines that interact because they are aware enough of the context in which they operate to be able to do so."[18] Since then numerous scholars have engaged not only with how the geoweb became interleaved with everyday life but also with what the consequences of this process have been (and might be) for urban space.[19]

As Thrift and French predicted, the emergence of the geoweb was largely a result of broad changes in the ways that users interacted with digital information through the Internet. Over time digital interaction evolved from being dominated by one-way informational transactions to being facilitative of omnidirectional flows of information. The early one-way user interactions with technologies of the late 1990s and early 2000s were (unsurprisingly) primarily consumptive. Computers with Internet connections were used, in other words, to find and consume information, and interactions with these technologies were initiated by explicit actions made by the user—to search the Internet for information, a user would query a search engine and the search engine would respond to the query by returning a string of data. Countless changes to how users engaged with digital technologies occurred throughout the first decade of the 2000s, and these changes had the net result of more effectively blending digital technologies into the everyday lives of users. By mid-decade, for instance, Internet users were more actively participating in the production of information by virtue of greatly expanded "Web 2.0" infrastructure (e.g., interactive web pages and relational web-based databases).[20] Flows of information could by mid-decade move in any direction (user to user, user to database, database to user, etc.), unrestricted by software or hardware limitations. By the end of the decade not only had mobile locative technologies become a more common feature in the urban landscape, but they were beginning to exhibit early signs of context awareness that subsequently has enabled them to anticipate and preempt users' explicit calls for information.[21] The contemporary landscape (mid-2010s) is one in which smart mobile technologies are a ubiquitous part of the urban landscape. From phones and tablets to wearables such as watches and eyeglasses, digital hardware and pervasive Internet connectivity have been increasingly woven into the fabric of everyday life for users across a wide socioeconomic spectrum. In an interesting contrast to the hardware and Internet connections of the late 1990s and first decade of the 2000s, affordability has been a key part of the wider diffusion of emergent technologies across urban space. Although the cost of some wearables, such as Google Glass, has (as of late 2014) been a limiting factor to their wider adoption, this is not the case with lower-cost wearables (such as smartwatches) and smartphones and tablet-style devices, particularly those that are subsidized by mobile phone and data services.

Scholars have consistently found that as mobile, locative, and pervasively Internet-connected devices become more ubiquitous, they begin to alter the ways that individuals perceive and experience everyday life in urban space.[22] Most striking has been the meshing of web-based geospatial data (the geoweb) with material socio-spatial practice. It is not at all uncommon, for example, to depend entirely on information provided by services such as Google Maps, Yelp, Foursquare, or some analogous application when navigating the city. Especially in unfamiliar places the mobile device is alternately a constant form of geospatial assurance and an augmented form of spatial cognition; though users might not be locals, their devices can tap a virtually limitless wealth of *seemingly* local knowledge on endless topics such as the locations and reviews of restaurants and coffee shops, home prices and rental rates, crime rates and school scores, walkability, traffic patterns, and park ratings. Notable is that the information consumed via mobile device is likely also the product of user production via mobile device—the pervasive Internet can in other words be alternately characterized as a repository for the producers of user-generated information (or volunteered geographic information, VGI) or as a destination for seekers (consumers) of web-based geospatial information. There are shockingly few obstacles to participation as information producer and even fewer obstacles for those who only consume, arguably leading to the emergence of a digital spatial imaginary that masks the uncertain (and biased) nature of its production in the authenticity of high-profile data services provided by well-known names such as Google, Yahoo, Microsoft, and Apple. Thus, in combination with the rise of context-aware anticipatory technologies, there is great potential for the experience of everyday life in urban space to subtly reflect the perceptions, tastes, and biases of other (unknown and anonymous) technology users. There is also considerable potential for technologies themselves to reproduce biased and/or inaccurate representations of space. This can be illustrated no more clearly than in the case of the Ghetto Tracker application, which was developed with the explicit aim of enabling users to characterize and avoid "unsafe" urban spaces (so-called ghettos). In reality, however, the Ghetto Tracker app simply enabled users to use locative web-based technology to demarcate poor and racially diverse neighborhoods, ultimately so that other users would be spared the inconvenience and/or discomfort of traveling through them.[23] As should be evident, the digital/socio-technical landscape of contemporary urban space is vastly more complicated than that of the late 1990s. To reiterate, there is no longer serious debate as to whether access to baseline digital technologies and Internet services has been achieved in urban space. Literacy has indeed improved alongside the diffusion of these technologies. Instead several exceedingly more difficult sets of questions related to digital exclusion have become apparent.

Confronting New Dimensions of Digital Exclusion?

There are at least three broad categories of questions that might be adopted as we begin to address new dimensions of digital exclusion in urban space:

1. How are new technologies becoming embedded in everyday life in the city?
2. Who has become, or is becoming, marginalized by the diffusion of new technologies in urban space?
3. What are the socio-spatial consequences of the embeddedness of new technologies in everyday life in the city?

By suggesting these categories I do not diminish the complexity of contemporary digital exclusion but instead underscore the necessity of identifying those moments at which we (as scholars, practitioners, policy makers, and activists) are best situated to act. In a recent piece Warf has highlighted the complexity of contemporary digital exclusion by suggesting that "the digital divide is multi-dimensional and cannot be reduced to dichotomous measurements, but must instead be viewed as a continuum measured using several variables."[24] Referring to a differentiation of "internet functionality, namely, what users do with cyberspace," he also offers that increasingly "the degree to which digital technologies mediate everyday life can lead to qualitatively different lifestyles between those who have access and those who do not."[25] Yet the word *access* here does not refer solely to the dichotomous relationship between user and computer that was characteristic of the first-order divide (though Warf does detail ongoing limitations to access among several subsets of the general population). Access is instead understood to exist on a continuum. A user, for instance, may own a computer with an Internet connection or a mobile device with a data plan, but how she or he interacts with those devices and subsequently uses those devices to interact with digital information is invariably connected to a broader notion of access to the digital realm. Digital exclusion is therefore both a characteristic and consequence of this continuum. Characteristically representing the socio-cultural, economic, and socio-technical limits to a user's full participation in digital technologies, exclusion from the full range of possibilities in everyday life can also be understood as a consequence of these limits. Thus, as we endeavor to break this increasingly complex and ongoing cycle of digital exclusion that began nearly two decades ago with the first-order divide, it is useful to conceive of contemporary dimensions of exclusion using these three categories of questions.

First and most simply: how are new technologies becoming embedded in the everyday social, economic, and spatial flows of life in contemporary cities? There has been considerable speculative and empirical work dedicated to this question since the middle of the decade, and with the pace of technological change quickening we can expect much more.[26] It is important, however, that new research be intensively oriented toward the embeddedness and facilitative characteristics of new digital technologies, as there is much potential for unexpected outcomes such as marginalization, exploitation, and the loss of privacy (i.e., surveillance) to become normalized by research that overstates the social or economic "good" of new technology. As new technologies become more pervasive—always on, locative, context aware, and Internet connected—their embeddedness in everyday life is less obtrusive. Wearable technologies in particular are designed to preempt explicit actions by the user by learning how, when, why, and where to initiate the transfer of information. Signals from our devices are in other words making their way unprompted from device to user as a consequence of the algorithmic logic embedded in the digital mind of the phone, watch, glasses, or other device. Human interaction with digital technology bears very little resemblance to interaction of the late 1990s and early 2000s, and so the ways that we frame research on digital access or differential patterns of usage must not only acknowledge the loosely organized chaos of contemporary digital interaction but also provide adequate mechanisms for understanding the many and varied modes of interaction.

Second, and most closely related to earlier work on the first- and second-order divides, we must take care in new research to explore critically who has become (or is most likely to become) marginalized (or excluded) by the diffusion of new technologies across urban space. Scholars in multiple disciplines have engaged with this question since the early 2000s, most recently resulting in efforts to understand not only who (in broad terms) the participants in digital technology tend to be but also how participation in digital technology has become increasingly differentiated.[27] No longer, for example, can participation be classified as either "on the Internet" or "not on the Internet." Yet, and as noted above, though mobile technologies have become near ubiquitous features in the urban landscape, there is significant differentiation among users when considering *how* those devices are embedded and deployed in everyday life. There have been indications that patterns of differentiation of technological participation adhere to the socioeconomic and cultural fissures that have long been distinct in urban space.[28] Approaches to conducting empirical research into the severity and/or rigidity of race- and class-based patterns of digital exclusion in the contemporary landscape are, however, less clear. New research programs must recognize the multiple dimensions of digital embeddedness and then explore socioeconomic and cultural exclusion, marginalization, and inequality by focusing on a particular manifestation of digital embeddedness in everyday life. When asking these types of questions, there is adequate room to engage at any scale—from coarse socioeconomic research on, for instance, the role of augmented reality applications in the daily lives of urban residents across multiple metropolitan areas to finer-grained qualitative or participatory work on patterns of, for instance, geosocial media participation among social groups within particular urban neighborhoods. In either case the outcomes should generate new evidence to advance theory not only on *who* (broadly speaking) tends to participate in new digital technologies but also *how* and *why* participation occurs and ultimately if there is an emergent geography of participation.

Finally, given both the ways that new technologies are woven into everyday life in the city and the patterns of marginalization that are becoming apparent, what are the socio-spatial consequences of new technologies for contemporary cities? Research in this area has been, and increasingly should be, driven largely by the intersection and manifestation in urban space of the first two sets of questions that I have posed above: How are technologies becoming embedded? And, who is/has been marginalized in the process?[29] As such, research in this area engages alternately with the socio-spatial experience of urban residents, as well as the ongoing processes by which urban spaces are produced. Scholars in this area have the opportunity to draw into their work a wealth of literature that has focused on the production of space, that engages with the complexities of socio-spatial relationality, and that contributes to an ongoing understanding of the geoweb in everyday life.[30] The outcomes (consequences?) of layering a new socio-technological landscape onto the existing urban fabric are profound. Through subtle (technologically mediated) adjustments to how, why, when, and where social, cultural, and economic activities are manifest in urban space, the material reality of the city is reproduced (albeit gradually) to better "fit" the digital landscape. What the digitally mediated reproduction of urban space means for the socio-spatial fabric of the city remains to be seen and is therefore an area ripe for scholarly inquiry. How, for instance, will the diffusion of technologies across urban space impact the social and economic flows in socioeconomically distressed urban neighborhoods? Or are increasingly surveilled public spaces facilitative of the same range of social, cultural, and economic activities? Urban space invariably is affected by digital mediation, and the range of inquiries into these effects is vast.

As we move forward, our collective ability to overcome new modes of digital exclusion hinges on rethinking how exclusion is manifest in everyday life. I have suggested three areas within which to focus our efforts, but I also emphasize that there remains significant uncertainty over the practices to undertake in order to enact change. Reliance on classical practices such as funding community technology centers or establishing regional public WiFi networks are less likely to have a measurable effect on the contemporary modes of digital exclusion. Instead action and inquiry must scale down to the particular practices, interactions, and outcomes of digital activities in urban space in order to identify effective moments of intervention. This means that one-size-fits-all types of interventions are far less likely to succeed than context-specific strategies that recognize the unique social, cultural, economic, and spatial characteristics of particular places. And of course we are not approaching technological equilibrium;

there are untold years of socio-technical change ahead of us, and with each iteration of change will emerge a host of new and unexpected modes of digital exclusion.

Notes

1. Brown, Barram, and Irving, *Falling through the Net.*
2. Brown, Barram, and Irving, *Falling through the Net.*
3. Brown, Barram, and Irving, *Falling through the Net.*
4. Kruger, *Access Denied*, 7.
5. Cotten et al., "Using Affordable Technology"; Kvasny and Keil, "Challenges of Redressing"; Kvasny, "Cultural (Re)production of Digital Inequality"; Servon, *Bridging the Digital Divide.*
6. Federal Communications Commission (FCC), *Deployment of Advanced Telecommunications Capability.*
7. Campbell and Kwak, "Mobile Communication and Civic Life"; Sylvester and McGlynn, "Digital Divide"; Haythornthwaite and Kendall, "Internet and Community."
8. Mesch and Talmud, "Internet Connectivity."
9. Lindsay, "Employability."
10. Green and Goetting, *Mobilizing Communities.*
11. On low digital literacy paralleling other socioeconomic challenges, see, for instance, Hampton, "Internet Use"; and Lindsay, "Employability."
12. Hargittai and Hinnant, "Digital Inequality."
13. Meyen et al., "Internet in Everyday Life."
14. See, for instance, Bobkowski and Smith, "Social Media Divide"; Blank, "Who Creates Content?"; and Mossberger et al., "Unraveling Different Barriers."
15. S. Graham, "Software-Sorted Geographies"; Crang, Crosbie, and Graham, "Technology, Time-Space"; Dodge, Kitchin, and Zook, "How Does Software Make Space?"; Williams and Dourish, "Imagining the City"; Kelley, "Semantic Production of Space"; Wilson, "Location-Based Services"; Brighenti, "New Media and Urban Motilities"; Kelley, "Urban Experience"; Kitchin and Dodge, *Code/Space.*
16. Hudson-Smith et al., "Mapping for the Masses."
17. See, for instance, Elwood, "Geographic Information Science" (parts I and II); Haklay, Singleton, and Parker, "Web Mapping 2.0"; Hudson-Smith et al., "Neo-Geography and Web 2.0"; Goodchild, "Citizens as Sensors."
18. Thrift and French, "Automatic Production of Space," 315–18.

19. See, for instance, Kitchin and Dodge, "Code and the Transduction of Space"; Dodge, Kitchin, and Zook, "How Does Software Make Space?"; Kitchin and Dodge, *Code/Space*; S. Graham, "Software-Sorted Geographies"; Graham and Zook, "Visualizing Global Cyberscapes"; M. Graham, Zook, and Boulton, "Augmented Reality in Urban Places"; Crampton et al., "Beyond the Geotag"; Leszczynski, "On the Neo in Neography"; Wilson, "Data Matter(s)"; Wilson, "Location-Based Services"; and Thatcher, "Living on Fumes."

20. See Kelly, "We Are the Web"; and O'Reilly, "What Is Web 2.0."

21. Kinsley, "Futures in the Making"; Kelley, "Semantic Production of Space."

22. Wilson, "Location-Based Services"; Mervyn and Allen, "Sociospatial Context"; Kelley, "Urban Experience"; Wilken, "Mobilizing Place."

23. Narula, "Real Problem"; Thatcher, "Avoiding the Ghetto."

24. Warf, "Contemporary Digital Divides," 2.

25. Warf, "Contemporary Digital Divides," 2.

26. Dodge and Kitchin, "Software, Objects, and Home Space"; Kitchin and Dodge, *Code/Space*; Waller, "Information Systems 'In the Wild'"; Wilson, "Location-Based Services"; Kelley, "Semantic Production of Space."

27. DiMaggio et al., "Digital Inequality"; Hargittai and Hinnant, "Digital Inequality"; Broos and Roe, "Digital Divide"; Kvasny, "Cultural (Re)production of Inequality"; Mossberger et al., "Unravelling Different Barriers"; Kelley, "Emergent Urban Imaginaries." See, for instance, Blank, "Who Creates Content?"; Bobkowski and Smith, "Social Media Divide"; and Brandtzæg, Heim, and Karahasanović, "Understanding the New Digital Divide."

28. P. Miller, "Digital Divide"; Anthes, "Digital Divide Persists"; Pearce and Rice, "Digital Divides from Access to Activities."

29. Past research on these issues includes Kelley, "Semantic Production of Space"; Thatcher, "Living on Fumes"; Dodge, Kitchin, and Zook, "How Does Software Make Space?"; Dodge and Kitchin, "Crowdsourced Cartography"; Elwood, "Geographic Information Science" (parts I and II); and Wilson, "Location-Based Services."

30. On the complexities of socio-spatial relationality, see, for instance, Lefebvre, *Production of Space*; Massey, *Space, Place, and Gender*; Martin, "Constructing Place." See, for instance, Pierce, Martin, and Murphy, "Relational Place-Making"; Jones, "Phase Space"; and Healey, *Urban Complexity*.

PART 5

Talking across Borders

10 Bringing the Big Data of Climate Change Down to Human Scale

Citizen Sensors and Personalized Visualizations in Climate Communication

David Retchless

Geospatial data and technology have long been essential to climate change research and are increasingly used to improve public engagement with this important issue. In a pathbreaking article Arrhenius used global temperature and humidity data sets—including temperature data collected via shipborne observations from the voyage of the HMS *Challenger*—to show not only that increasing atmospheric carbon dioxide concentrations would warm the troposphere but also that this warming would be spatially variable.[1] Arrhenius presented his results in a table that showed expected temperature changes in ten-degree latitudinal bands, from 70° north to 60° south. Advances in sensing, networking, and computing technologies since Arrhenius's time have greatly increased the coverage, accessibility, speed, and complexity of geospatial data available to climate change researchers, while also changing the ways in which these researchers communicate their results. For example, the Intergovernmental Panel on Climate Change (IPCC) has used complex global computer models to project future temperature change, and its report assesses past temperature change using data from satellites, ships, ice cores, and terrestrial weather stations.[2] In a significant upgrade to Arrhenius's tables this IPCC report from 2013 uses color-filled isoline maps generated using geographic information systems (GIS) software and distributed over the Internet to communicate this temperature change information to scientists, policy makers, and interested members of the public.

The transition from Arrhenius's 1896 article to the 2013 IPCC report reflects many of the earliest advances of the geospatial revolution. Beginning with their widespread adoption in the 1970s and 1980s, satellite imagery and GIS technology have transformed the practice of the spatial sciences; with the growth of the Internet in the 1990s and broadband technologies into the 2000s, access to geospatial data and visualizations through geoportals (such as the Earth Systems Grid for climate data), assessment reports, and other media has rapidly expanded.[3] Much has already been written about the connections between these foundational technologies of the geospatial revolution and climate change. For example, a recent edited volume, *Geospatial Technologies and Climate Change*, covers both the specific types of geospatial technologies that have supported climate change research in recent decades (e.g., remote sensing, GIS, the Global Positioning System, and global climate models), and the role of these technologies in impact assessment and adaptation and mitigation planning.[4] However, with the advent of the modern geoweb— including geotagging, crowdsourcing, mash-ups, data mining, and mobile applications (apps) and sensors[5]—the big data of global satellites and climate models is increasingly supplemented by small data approaches that improve the accessibility and personal and policy relevance of climate data. This growing emphasis on small data approaches suggests that the relationship between climate change research and geospatial technology is evolving, with significant implications for both the generation and communication of climate change knowledge.

While both facets of this evolving relationship warrant additional research, this chapter focuses on the second: the interplay between these new geospatial technologies and climate change communication and engagement. Communication is defined broadly. It includes the traditional, top-down communication from climate scientists and other experts to a lay public.[6] But it also includes bi- and multidirectional communications among scientists, publics, and policy makers, which—furthered by the technologies of the geoweb that translate the big data of climate change to human scales—support citizen science and soften (without dissolving) the boundaries of scientific (and cartographic) authority.[7] Following W. Lin and drawing on Haklay, Singleton, and Parker, as well as Elwood and Leszczynski, I define the geoweb as the merging of geographic information with other web-based content, including the practices and tools that support this merger and the media that result from it.[8] As W. Lin describes, the practices and products of the geoweb (including volunteered geographic information and the tools for sharing it, such as OpenStreetMap) provide a platform for tying particular, geolocated information into larger, issue-based social networks, effectively bridging local and regional scales and lowering geospatial barriers to networking.[9] In this way the technologies of the geoweb may hold potential to tie individual experience of local climate drivers and impacts to the issue of global climate change through networks of similarly concerned and curious individuals. While data aggregated through such networks may be amenable to big data analyses, from a communications perspective these networks' potential for sharing small data about global issues (such as climate change) may be more valuable: as Kitchin and Lauriault argue, small data "can focus on specific cases and tell individual, nuanced and contextual stories" that connect with audiences in a way that global data sets cannot.[10]

As discussed below, citizen use of mobile apps and sensors may be one such means of using the geoweb and small data to diversify the directionality of climate change communication; feedback from map and app users may similarly enable highly personalized visualizations of the geography of climate change, lessening dependence on the one-size-fits-all loading dock communication model. This chapter considers the potential of these new approaches to increase citizen engagement—defined here as understanding, interest, and the motivation to take action—with climate change.[11] The chapter begins with a review of existing challenges in communicating climate change to increase engagement, with a focus on how geospatial tools have shaped these challenges. It then explores how the two approaches to climate change communication introduced above—citizen sensors and personalized visualizations—may address these challenges.

Challenges

Much of our knowledge about the "nature and characteristics of climate change" comes through "massive," remotely sensed, global data sets.[12] These geospatial technologies provide a big data lens that makes visible the phenomena of climate change but that may also limit our view in ways that frustrate engagement.

Authors have identified a long list of obstacles to fostering engagement with climate change among the public and policy makers.[13] In addition to describing how individual knowledge, beliefs, and feelings can interact with social structures and institutions to create a hostile environment for engagement, these authors also identify at least three ways in which the phenomenon of climate change—as viewed through the lens of geospatial technology—is itself challenging to understanding, interest, and action. First, the huge spatio-temporal scales involved may lead audiences either to discount climate change impacts as a problem for the far future or distant places or to despair, believing that such impacts are a global problem too big for individual action to address.[14] Second, because the causes of climate change—such as increasing atmospheric greenhouse gas concentrations—cannot be seen and can be difficult to understand, audiences often find it difficult to connect their abstract understanding of these causes with their more emotionally rich experience of everyday weather and extreme events.[15] Third, the great complexity of the climate system and the uncertainties inherent in using models and other technologies to understand it may frustrate some audiences and encourage skepticism and mistrust in others.[16] The relationship between each of these three challenges and geospatial technology is explored in more detail below.

Mismatched Scales

Many of the challenges of communicating climate change have roots in the broad spatial and temporal extent of the geospatial data and analyses that ground research in this field. Depictions of climate change as a global or national problem that will play out over decades to centuries may encourage audiences to discount its personal importance, leading them to believe that any negative impacts will primarily be felt in the distant future, by people who live far away, or by nonhuman nature.[17] As I discussed in 2014, several authors have responded to this disconnect by recommending a shift from a global climate change discourse toward a local discourse that shows how climate is embedded in communities and how changing climates will affect people's daily lives—in short, how climate change is locally relevant and personally meaningful.[18]

Such downscaling efforts face significant technical obstacles, since climate projections at local to regional scales tend to be much less certain than global scale projections.[19] One solution to this problem has been to use GIS and 3-D modeling software to create visualizations of multiple climatic and socioeconomic scenarios for local communities.[20] Such visualizations can show engaging, photo-realistic representations of how changes in local sea level and ecosystems in response to global climate change may combine with development choices to reshape human landscapes; S. Sheppard et al. describe the application of such visualizations to waterfront, urban, and hillside communities in Vancouver, British Columbia.[21] While such visualizations cannot show all possible future situations, they can present scenarios of local change that acknowledge uncertainty while also helping communities and their residents to connect global climate change to daily life in their community.[22]

All Context, No Feeling

Because neither the causes nor the immediate effects of climate change are directly visible, our understanding of climate change is necessarily mediated by remote sensing and GIS technologies that observe, aggregate, and translate this information into forms—such as tables, graphs, and maps—that are easier to perceive but lack the immediacy of direct experience.[23] Trends in global concentrations of carbon dioxide and other greenhouse gases thus become visible when measured at monitoring stations such as Mauna Loa Observatory, and climate change signals emerge from the noise of daily weather when data or model output are aggregated over large areas and time periods.[24] While these abstractions make climate change visible, they are much less likely to generate a visceral reaction than are hazards that are experienced directly.[25] When combined with availability heuristics, this lack of direct climate change experience may explain why nonexperts often base their assessment of climatic change primarily on experiences that are easier to call to mind, such as weather experienced over the past year.[26] Such conflations of weather and climate can further obscure the link between cause and effect: the science may show steadily increasing atmospheric concentrations of greenhouse gases over a period of several decades, while direct experience of a cool summer may lead one to erroneously conclude that global temperatures are not rising in response to these increased concentrations, as scientists predict.

Climate change communicators thus face two related and potentially conflicting objectives: to place personal experience in the context of global change and to enliven and make more immediate the presentation of global climate change observations and projections. Sheppard suggests that the responsible use of realistic 3-D visualizations and flybys showing dramatic views of plausible climate change impacts (such as "forest die-back or the effects of extreme storms") can address the second objective by making climate change projections more immediate.[27] It is not clear, however, that such use of "permissible drama" can address the other objective: effectively connecting everyday experience of less dramatic environmental changes to global climate change.[28] As discussed in the "Opportunities" section below, facilitating the creation and sharing of small data stories may make new geoweb-based technologies more successful in engendering an emotional connection with climate change while simultaneously contextualizing personal experience.

It's Complicated

The complexity of the climate system—and the models and technologies used to understand it—can also complicate the communication of this information to the public, especially when dealing with uncertainties in climate change projections.[29] Projections of climate change depend on multiple, interacting uncertainties. These uncertainties derive from many different sources, including future emissions trajectories, natural variability in the climate system, and uncertainties in the climate models themselves.[30] Each of these sources of uncertainty may also entail several different types of unknowns; MacEachren et al. identify nine such "components" of uncertainty, all of which could be applied to climate change projections.[31]

Communicators must decide whether and how to show these different sources and types of uncertainty in the maps and other tools they use to make climate change visible to their audiences. Assessment report authors often do choose to show some types and sources of uncertainty; for example, their maps of changes in temperature or precipitation will frequently use coincident bivariate techniques to show uncertainty (usually the precision of climate model output) as an overlay or mask on top of the thematic map data.[32] Including uncertainty information in this way need not interfere with other map-reading tasks and may help improve decisions.[33] Although some authors have cautioned that uncertainty may be used to bar or delay action on controversial issues such as climate change by encouraging skepticism or mistrust, others have stressed the importance of including uncertainty in depictions of climate change used in decision-making contexts, where outliers can carry important policy implications.[34] While these considerations point to a need for communicating more uncertainty information, designers of maps and other geospatial visualizations of climate change also face limits to the number of attributes that can be shown without cluttering the display, suggesting that additional uncertainty information should be provided only where and when needed.[35]

Despite this support for including depictions of uncertainty in climate change communications, important questions remain. Research continues to explore and evaluate new and existing combinations of visual variables for mapping uncertainty, both in general and for climate change in particular.[36] The effect of individual differences on perceptions of uncertainty and risk is also an area of active research, with prior hazard beliefs and experience, affect and emotion, numeracy, gender, and domain expertise all having been shown to influence risk perception.[37] Based on this research, visualizations of climate change could perhaps be customized to vary (or omit) the display of uncertainties for different audiences. As discussed below, geoweb-based technologies may create new opportunities for implementing such customized displays.

Opportunities

By reshaping the geospatial lens through which the public views climate change, emergent geoweb and small data approaches may help address these challenges. In her review of the literature on uses of the geoweb, Elwood describes how geoweb research has been focusing on the use of visualization and networking technologies to promote awareness and activism.[38] Most of this research has focused on geopolitics, power, human rights, and violence (perhaps reflecting the researchers' strong connections to critical and feminist GIS and cartography).[39] However, the geoweb also seems well suited to promoting awareness of environmental issues such as climate change.

Elwood identifies two ways in which the geoweb may enhance awareness and activism: the long-distance, one-to-many sharing of crowd-sourced, geotagged information, as well as the "overhead large-area views with more immediate 'on the ground' views in virtual globes."[40] These parallel Moser's identification of both the role for dialogic communication and the effective use of visualization as future research directions for climate change communication.[41] They also match two existing uses of the geoweb—for citizen sensors and personalized visualizations—for climate change communication. The two short sections that follow briefly consider the potential of these two uses of the geoweb as means of enhancing public engagement with climate change.

Citizen Sensors

Goodchild introduced the term "volunteered geographic information" (VGI) to describe the then-emerging phenomenon of "the widespread engagement of large numbers of private citizens, often with little in the way of formal qualifications, in the creation of geographic information."[42] In this same article he identified the use of citizen sensors for citizen science as one of the primary applications of VGI. Citizen science has a long history in the weather and climate communities, predating the geoweb by more than a century. As part of the U.S. National Weather Service's Cooperative Observer Program (COOP), volunteers have been collecting daily, cite-specific weather data since the creation of the program in 1890; however, the practice of citizens recording daily weather stretches back even further, to the colonial period.[43] Citizen phenological observations have a similarly lengthy record. Henry David Thoreau marked the changing of the seasons in Concord, Massachusetts, during the 1850s with meticulous observations of plants' flowering times; comparison with modern observations has revealed how climate change is altering the local ecology.[44] Since 1901 the U.S. National Audubon Society has enlisted thousands of volunteers from locations across the United States and Canada each December for the Christmas Bird Count. Researchers have recently used these observations to suggest shifts in birds' winter ranges as the climate warms.[45]

Building on this long tradition of citizen-driven climate science, the technologies of the geoweb are enabling citizens to share, aggregate, and retrieve observations from around the globe faster and more easily than ever before, allowing them to assemble small data stories that tie local and regional observations to global change. GPS and the responsive design of Web 2.0 technologies facilitate sharing of geotagged observations on web maps and virtual globes, while mobile phones outfitted with GPS, cameras, and other sensors simplify the process of collecting "scientifically relevant data as part of [one's] daily routine."[46] Applying these technologies to the gathering of weather and climate data, the GLOBE Program enlists students and teachers worldwide to contribute weather, landcover, soil, and hydrology data through their school.[47] Similar to the National Weather Service's COOP but open to schoolchildren and being global in scope, GLOBE makes all data that students collect available via a Google Maps–based GIS, with an option to view data on the Google Earth virtual globe. For phenology data there are various mobile apps, such as Budburst for plant watching and eBird/BirdLog for bird counts, that make it easy for citizen scientists to quickly geotag and upload their observations, which may include not only text but also smartphone photos with timestamps and GPS coordinates.[48] These services also provide online, interactive maps that allow app users to view the submissions of other observers, whether nearby or far away.

In addition to lowering the barriers of entry for those who are interested in participating in citizen science, the proliferation of cameras, GPS units, and other sensors on cell phones and other mobile devices has also created "accidental" geographic information: information that may be volunteered for one purpose but used by researchers for another. For example, a vacationer might upload time- and location-stamped outdoor photos to social media or photo-sharing websites to share with friends, but researchers might download and analyze these photos to learn more about the weather at the place and time they were taken. Hyvärinen and Saltikoff have demonstrated that such a data-mining approach could be used with photos from Flickr to validate radar-indicated hailstorm cells with on-the-ground observations.[49] While this suggests that such techniques for data-mining social media may hold potential for learning about weather and climate, researchers should also consider whether it is ethical to study data not explicitly shared for research purposes.[50] More broadly, researchers should also attend to the ethical implications of other aspects of consent involved in the crowdsourcing of geolocated weather and climate information, including whether consent was offered via an opt-in or opt-out mechanism and whether the publication of such geolocated information online implies consent to use it in research, even when such use may have significant privacy implications.[51]

Two key components of voluntary participation in geoweb-enhanced citizen science may be particularly applicable to the climate change communication challenges identified above: its ability to span "multiple spatial, temporal, and social scales" and its promotion of effective and multidirectional communication among participants and organizers.[52] By providing tools to facilitate citizen scientists' efforts to look for evidence of global climate change in their local communities, the technologies of the geoweb can help these citizens connect their everyday experiences to larger processes of global change. Unlike the top-down communication of climate change information through channels such as assessment reports, citizen science apps can engage the public in the production of new climate change knowledge.[53] By placing their observations on virtual globes or interactive online maps that also show observations from other scientists (both citizen and traditional), these newly engaged citizen scientists can gain the context needed to understand how their observations fit within larger global trends.

Personalized Visualizations

In addition to supporting citizen sensors and science, the technologies of the geoweb have also encouraged the development of visualizations that are personalized to suit users' interests and locations. Users can deploy web-mapping APIs to create interactive maps and mash-ups that are customized to meet their needs and then use social platforms to share these custom maps and make them easily discoverable.[54] And with pan-and-zoom interfaces users can, with GPS, zoom in on these maps to show only nearby areas or zoom out and explore other areas of interest.[55] In a possible reflection of nonprofit and government organizations' recognition of the power that pan-and-zoom, details-on-demand tools for geographic exploration hold for communicating seemingly distant or inaccessible issues like climate change, nonprofits and government agencies have now contributed more than fifty layers to the climate change section of Google Earth's outreach showcase.[56] In announcing its climate data initiative, the federal government has also formally recognized the need to translate the big data of global climate into the small data of local simulations, issuing challenges to "researchers and developers" for the development of new, interactive visualizations that use these data "to educate the public about the vulnerability of their own communities."[57]

Personalized visualizations may also help to foster engagement, build trust, and encourage learning about complex environmental problems such as climate change—particularly when combined with VGI and other practices that blur the boundaries between creators and consumers of both geographic data and scientific knowledge. In an example of how platforms for collecting and visualizing VGI may improve environmental engagement and awareness, W. Lin describes how the website FindingChina—a Google

Maps mash-up with VGI contributions highlighting Chinese environmental problems—has created "forums for public engagement" where users can share geolocated information about local environmental problems, explore problems in other locations of interest, and build communities around addressing these problems.[58] When such platforms for collecting and visualizing VGI are used to facilitate collaborations between the public and scientific experts, they may also promote public understanding of and trust in these experts' assessments of complex environmental hazards. For example, Young et al. describe how a web-based U.S. Geographical Survey platform for gathering and mapping crowdsourced reports of the intensity of shaking during earthquakes may help those making a report better understand their future earthquake risk.[59] By showing where their experience falls on a scientific earthquake intensity scale and placing this experience in the context of other users' reports, the USGS tool helps reporters understand what the numbers on the intensity scale mean and how damage and other impacts vary with changes in intensity. Haklay suggests that such collaborations can break down barriers between citizens and scientists, potentially building public trust and the public's understanding of scientific endeavors.[60]

These examples suggest that personalized visualizations can address the challenges of climate change communication in several ways. By allowing users to zoom in on areas that are nearby or of interest, personalized visualizations may make clear how an abstract global problem such as climate change will affect people and places that are local, familiar, and important to the user. Additionally, visualizations that allow users to zoom out from local to regional and global views can allow users to contextualize these impacts. By allowing users to choose from among several possible futures (e.g., those with different levels of greenhouse gas emissions, temperature change, or sea level rise), visualizations that can be personalized to display multiple climate change scenarios may also help to communicate uncertainty and raise awareness of worst-case scenarios. And, as discussed above, when combined with VGI and used in collaboration with scientific experts, such visualizations may also encourage trust in and understanding of climate science.

Evidence suggests that including uncertainty information on geospatial displays does not hinder (and may improve) comprehension of map content.[61] It may nevertheless in some cases be desirable to customize displays based on individual differences that have been shown to affect perceptions of uncertainty and risk (see review from the "Challenges" section, above). While many climate change visualizations (especially those for mobile apps) already use GPS to automatically personalize their display based on user location, automatic personalization based on user interests or other individual differences may be feasible but is not yet popular. Senaratne et al. suggest that surveys of potential users could gauge their expertise and that the display of uncertainty could then be adjusted accordingly.[62] This approach could help address the challenge of communicating climate change uncertainty to audiences with different levels of comfort and familiarity with the complexities of climate science. However, additional research would be needed to determine which user characteristics best fit which depictions of uncertainty. Rather than surveying users directly, developers of visualizations could also use behavioral targeting to infer users' interests and level of expertise from their web-browsing habits and then customize their visualization accordingly. Such an approach would likely be controversial, since behavioral targeting is an online advertising technique that can raise privacy concerns.[63] As with the privacy concerns inherent in crowdsourcing and data mining of geolocated information, developers of web pages or apps that use behavioral targeting should be mindful of obtaining consent for the collection and use of personal data.[64]

Conclusions

By facilitating the creation and sharing of climate observations, placing observations in the context of other observations and climate data, enabling exploration of such data across local to global scales, and bridging the divide between big and small climate data, the technologies of the geoweb appear to hold significant promise for increasing climate change engagement. Several aspects of the use of citizen sensors and personalized visualizations to communicate climate change may warrant further study. For citizen science and other uses of VGI, organizers and participants will need to negotiate control over data access and privacy. This concern may become particularly pressing if citizen science moves from its current siloed model of multiple, independent projects with little interaction to the integrated model proposed by Newman et al., which would support sharing of data and results across multiple projects.[65] For visualizations such as virtual globes researchers should carefully consider how highly realistic and localized depictions of climate change may affect users' belief in their accuracy. Because trust in the source of climate change information is a key factor in how it is received, future research should consider how presenting climate change information using such virtual globes (or other, similar interactive geovisualizations) affects user beliefs about the trustworthiness of the source and the veracity of the data.[66] As these visualizations move from virtual globes to augmented reality applications, the increased visibility of climate hazards may raise important legal and social justice questions. Goodchild relates how a plan to paint a blue line across the streets of Santa Barbara, California, showing the future shoreline after a seven-meter rise in sea level, was defeated by developers and homeowners who feared that the line would negatively affect their property values.[67] This example suggests that while believable and highly visible depictions of climate impacts may succeed in engaging audiences, they may also become foci for conversation and conflict in local communities as residents negotiate the meaning of public and private spaces being marked as "vulnerable to climate change." Additional research is needed to better understand how communities will react to such spatially explicit and highly visible depictions of climate impacts.

Notes

1. Arrhenius, "On the Influence of Carbonic Acid."
2. Stocker et al., *Climate Change 2013*.
3. Foresman, *History of Geographic Information Systems*; Maguire and Longley, "Emergence of Geoportals"; Bernholdt et al., "Earth System Grid."
4. Sundaresan et al., *Geospatial Technologies*.
5. Haklay, Singleton, and Parker, "Web Mapping 2.0"; Elwood, "Geographic Information Science" (parts I and II).

6. This approach has been justly criticized for following the "loading dock" (Beller-Simms and Ingram, *Decision-Support Experiments*) and "information deficit" models (Sturgis and Allum, "Science in Society"; Pidgeon and Fischhoff, "Role of Social and Decision Sciences"), which assume that experts know what information the laity needs, that delivering this information is sufficient to ensure understanding, and that understanding this information is sufficient to engender agreement and engagement.

7. Goodchild, "Citizens as Sensors."

8. W. Lin, "Volunteered Geographic Information"; Haklay, Singleton, and Parker, "Web Mapping 2.0"; Elwood and Leszczynski, "New Spatial Media."

9. W. Lin, "Volunteered Geographic Information."

10. Kitchin and Lauriault, "Small Data in the Era of Big Data."

11. Lorenzoni, Nicholson-Cole, and Whitmarsh, "Barriers Perceived"; Ockwell, Whitmarsh, and O'Neill, "Reorienting Climate Change Communication."

12. Sundaresan et al., *Geospatial Technologies*, vii.

13. Pidgeon and Fischhoff, "Role of Social and Decision Sciences"; Moser, "Communicating Climate Change"; Nisbet, "Communicating Climate Change"; Ockwell, Whitmarsh, and O'Neill, "Reorienting Climate Change Communication"; Lorenzoni, Nicholson-Cole, and Whitmarsh, "Barriers Perceived."

14. Pidgeon and Fischhoff, "Role of Social and Decision Sciences"; Moser, "Communicating Climate Change"; Ockwell, Whitmarsh, and O'Neill, "Reorienting Climate Change Communication"; Lorenzoni, Nicholson-Cole, and Whitmarsh, "Barriers Perceived."

15. Pidgeon and Fischhoff, "Role of Social and Decision Sciences"; Moser, "Communicating Climate Change"; Nisbet, "Communicating Climate Change"; Ockwell, Whitmarsh, and O'Neill, "Reorienting Climate Change Communication."

16. Pidgeon and Fischhoff, "Role of Social and Decision Sciences"; Moser, "Communicating Climate Change"; Nisbet, "Communicating Climate Change"; Ockwell, Whitmarsh, and O'Neill, "Reorienting Climate Change Communication"; Lorenzoni, Nicholson-Cole, and Whitmarsh, "Barriers Perceived."

17. See, e.g., Stocker et al., *Climate Change 2013*; Melillo, Richmond, and Yohe, *Climate Change Impacts in the United States*; Nicholson-Cole, "Representing Climate Change Futures"; Leiserowitz, "Communicating the Risks of Global Warming"; Lorenzoni, Nicholson-Cole, and Whitmarsh, "Barriers Perceived"; Swim et al., *Psychology and Global Climate Change*.

18. Retchless, "Communicating Climate Change." See also work by perceptions specialists, such as Slocum, "Polar Bears and Energy-Efficient Lightbulbs"; Leiserowitz, "Communicating the Risks of Global Warming"; and Hulme, "Geographical

Work at the Boundaries of Climate Change," as well as work by geographers and other global change scientists on the regional impacts of climate change, such as the Global Change in Local Places project of the Association of American Geographers in 2003.

19. Hawkins and Sutton, "Potential to Narrow Uncertainty."
20. S. Sheppard, "Landscape Visualisation."
21. S. Sheppard et al., "Future Visioning of Local Climate Change."
22. A. Shaw et al., "Making Local Futures Tangible"; Sheppard et al., "Future Visioning of Local Climate Change."
23. Moser, "Communicating Climate Change"; Dunwoody, "Challenge of Trying to Make a Difference."
24. Thoning, Tans, and Komhyr, "Atmospheric Carbon Dioxide at Mauna Loa Observatory"; Hawkins and Sutton, "Potential to Narrow Uncertainty."
25. Weber, "Experience-Based and Description-Based Perceptions."
26. Keller, Siegrist, and Gutscher, "Role of the Affect"; Howe et al., "Global Perceptions."
27. S. Sheppard, "Landscape Visualisation," 649.
28. S. Sheppard, "Landscape Visualisation," 649.
29. Pidgeon and Fischhoff, "Role of Social and Decision Sciences."
30. Hawkins and Sutton, "Potential to Narrow Uncertainty."
31. MacEachren et al., "Visualizing Geospatial Information Uncertainty."
32. See, e.g., Stocker et al., *Climate Change 2013*; Melillo, Richmond, and Yohe, *Climate Change Impacts*.
33. See, e.g., Viard, Caumon, and Lévy, "Adjacent versus Coincident"; Deitrick and Edsall, "Influence of Uncertainty"; Brickner et al., "Better Decision Making"; review from Harrower, "Representing Uncertainty."
34. Freudenburg, Gramling, and Davidson, "Scientific Certainty Argumentation"; Oreskes and Conway, *Merchants of Doubt*; Freudenburg and Muselli, "Reexamining Climate Change Debates"; Oppenheimer et al., "Limits of Consensus"; Brown and Wilby, "Alternate Approach"; Lewandowsky et al., "Scientific Uncertainty."
35. Lohrenz et al., "Model of Clutter."
36. See, e.g., Zuk and Carpendale, "Theoretical Analysis"; and Kaye, Hartley, and Hemming, "Mapping the Climate."
37. Kellens, Terpstra, and De Maeyer, "Perception and Communication"; Loewenstein et al., "Risk as Feelings"; Slovic et al., "Risk as Analysis"; Severtson and Myers, "Influence of Uncertain Map Features"; Maibach, Roser-Renouf, and Leiserowitz, *Global Warming's Six Americas*; Roth, "Impact of User Expertise."
38. Elwood, "Geographic Information Science" (parts I and II).

39. Elwood, "Geographic Information Science" (parts I and II).
40. Elwood, "Geographic Information Science" (part II), 404.
41. Moser, "Communicating Climate Change."
42. Goodchild, "Citizens as Sensors," 212.
43. Fiebrich, "History of Surface Weather Observations."
44. Primack and Miller-Rushing, "Uncovering, Collecting, and Analyzing."
45. Valiela and Bowen, "Shifts in Winter Distribution in Birds"; Niven et al., "Christmas Bird Count Provides Insights"; Sorte and Thompson, "Poleward Shifts in Winter Ranges."
46. Haklay, Singleton, and Parker, "Web Mapping 2.0"; E. Graham, Henderson, and Schloss, "Using Mobile Phones," 313.
47. GLOBE Program, https://www.globe.gov/; Goodchild, "Citizens as Sensors."
48. Budburst, http://budburst.org/gomobile; eBird/BirdLog, http://ebird.org/content/ebird/news/birdlog-your-cbc/.
49. Hyvärinen and Saltikoff, "Social Media as a Source of Meteorological Observations."
50. Elwood, "Volunteered Geographic Information."
51. F. Harvey, "To Volunteer or to Contribute"; Blatt, *Health, Science, and Place*.
52. Newman et al., "Future of Citizen Science," 298.
53. Haklay, "Citizen Science."
54. Haklay, Singleton, and Parker, "Web Mapping 2.0."
55. Shneiderman, "Eyes Have It."
56. Elwood, "Geographic Information Science" (parts I and II).
57. Podesta, "Climate Data Initiative."
58. W. Lin, "Volunteered Geographic Information," 959.
59. Young et al., *Transforming Earthquake Detection*.
60. Haklay, Singleton, and Parker, "Web Mapping 2.0."
61. Viard, Caumon, and Lévy, "Adjacent versus Coincident"; Deitrick and Edsall, "Influence of Uncertainty"; Brickner et al., "Better Decision Making."
62. Senaratne et al., "Usability of Spatio-Temporal Uncertainty Visualisation."
63. Berger, "Balancing Consumer Privacy."
64. Blatt, *Health, Science, and Place*.
65. Newman et al., "Future of Citizen Science."
66. Kahan, "Making Climate-Science Communication."
67. Goodchild, "Commentary: Whither VGI?"

11 Synergizing Geoweb and Digital Humanitarian Research

Ryan Burns

Humanitarian work is increasingly incorporating diverse sources of information and labor. This trend represents a shift that has significant repercussions for social, political, and economic relations. Exemplified by websites and communities such as Ushahidi, the Humanitarian OpenStreetMap Team, and the Standby Task Force, this development, commonly called "digital humanitarianism," sits at the juncture of new socio-technical practices, new epistemologies, and new institutional relationships.[1] Digital humanitarianism involves formal humanitarian organizations tasking a large, unrestricted number of laypeople, often volunteers, with work related to gathering, producing, processing, publicizing, and mapping humanitarian data. For example, the Standby Task Force community members and Ushahidi users might be involved in collecting public information from social media and journalism sources, translating and processing that information, and visualizing it in a cartographic product to be used by formal responding agencies.[2] The Humanitarian OpenStreetMap Team normally produces spatial data in crisis-affected areas by tracing new vector-based data over remotely sensed images of crisis areas, in the OpenStreetMap platform. These data can include infrastructure, expressions of need, or reports of security threats. Digital humanitarians contribute to the production, processing, and visualization of spatial data based on the assumption that these kinds of data, in their platforms, are needed by formal responders to more effectively deliver aid and that volunteers—usually scattered across the globe—are able to deliver such data. Recent research on digital humanitarianism has focused on understanding and processing social media and mobile phone records as additional sources of data.[3] Digital humanitarianism thus encapsulates elements of the geoweb, crowdsourcing, crisis mapping, big data, and mass collaboration.[4]

Both digital humanitarian researchers and practitioners are increasingly foregrounding big data in their work, yet through narrow analytic lenses. These nascent conversations are often technical in nature, for instance, in determining a quantitative measure of reliability for social media data. At other times they grapple with the policy potentials and hindrances associated with integrating big data into formal humanitarian and emergency management workflows. Researchers and practitioners are actively involved in advancing and advocating for these technologies. With this applied focus under way, recent research has begun theorizing big data digital humanitarianism as a set of socio-technical practices and relations, seeking synergies with critical data and technology studies.[5]

Digital humanitarianism uses spatial technologies to establish relations between geographically distant populations. Despite this inherent spatiality, to date geographers have researched little about the field. In this chapter I explore ways geographers interested in big data can engage the field of digital humanitarianism to build empirical and theoretical connections with geoweb research. My primary goal is to draw on theoretical frameworks within geoweb research in order to highlight resonances with digital humanitarianism. I illuminate connections between existing big data, digital humanitarianism, and geoweb research but am primarily concerned with new, pressing questions and key linkages with existing research. Research on spatial technologies' social and political transformations provides the most compelling tools for conceiving of digital humanitarianism as a set of socio-technical practices and political-economic relations.

I begin by reviewing current research on digital humanitarianism. While more research is taking place outside geography's disciplinary boundaries, geographers are uniquely well positioned to theorize the social, political, and economic impacts of digital humanitarianism. I then detail three connections between digital humanitarianism and geoweb research: digital humanitarianism's spatialities, political economy, and knowledge politics. I conclude by drawing out larger implications of these connections.

Current Research on Big Data Digital Humanitarianism

Extant research on digital humanitarianism is predominantly descriptive and technical in nature. Conversations have revolved around determining data accuracy and credibility, changing operational roles of emergency managers, solutions to technical problems, and streamlining its integration into policy and operations.[6] An oft-given explanation for the nonadoption of such technologies by the formal humanitarian sector is a lack of confidence in the reliability and actionability of digital humanitarian data.[7] That is, the formal humanitarian sector often sees digital humanitarian data as either low in quality or of a nature that cannot inform established workflows and practices. Big data factors prominently into these discussions as constitutive of new data sources and forms, with varying levels of accuracy.[8] To these ends, current research seeks to improve data accuracy or to assign numerical values representing the given accuracy of the data.[9] In many cases research explicitly seeks to control false or misleading social media data.[10]

The emergence of big data has instigated new roles and responsibilities for formal humanitarian actors.[11] In particular the role of emergency managers is shifting to include monitoring and utilizing social media in crises.[12] As increasing numbers of laypeople contribute to crowdsourced data production (either wittingly or without informed consent), emergency and humanitarian managers are working to incorporate larger and more diverse data sources, displacing the formal sector from the role of being primary producers of data.[13]

Digital humanitarian operations face technological obstacles due to changing social and geographic contexts of implementation, evolving socio-technical practices, and emergent software platforms. As digital humanitarian technologies are mobilized in diverse places around the world, they require unique approaches and understandings of technology use.[14] This means adapting tools to local contexts in order to foster wide adoption and efficacious uses.[15] Some research has focused on digital humanitarian interventions in political conflict zones.[16] However, digital humanitarian research is largely aspatial, with "solutions" to problems framed as universal rather than geographically contingent.

The narrow digital humanitarian research focus on technical descriptions can be partly explained by recent technological advances and the resulting shift in social uses of technologies. Digital humanitarians are developing new technologies to cope with big data, including evolving security threats, unmanned aerial vehicles (UAVs, including the class of UAVs commonly called drones), and automating work currently undertaken by contributors.[17]

Formal-sector adoption of big data digital humanitarian technologies has been uneven in two senses. First, some institutions advocate for their adoption far more than others, with early adopters, including the United Nations' Office for the Coordination of Humanitarian Affairs (OCHA), the U.S. Federal Emergency Management Agency (FEMA), and the American Red Cross, taking prominent roles.[18] Agencies tend to adopt technologies directly related to their existing workflows and practices, rather than seeking to shift their practices and workflows, a tendency that tempers adoption.[19] Another common reason given for a lack of widespread adoption is that existing policies and workflows hinder the changes required to efficiently utilize digital humanitarian technologies; in many cases emergency managers perceive policy hurdles to be insurmountable.[20] Some researchers have conducted analyses and provided recommendations regarding how to navigate legal and policy issues of big data digital humanitarian technologies.[21]

Geographic Research

Despite the spatial nature of humanitarianism and the rich body of geographic literature critiquing it, geographers have to date not contributed much to a theorization of digital humanitarianism. Early research has sought to position digital humanitarian technologies and practices alongside the geoweb. For instance, Goodchild and Glennon draw lines of resonance between social media, crisis mapping, and volunteered geographic information (VGI), suggesting that data produced in digital humanitarian technologies must undergo veracity tests similar to those used for traditional data.[22] Goodchild and Glennon are concerned primarily with this question of the *accuracy* of VGI, with accuracy being conceptualized as a measure of difference between an observed phenomenon's location on the earth and the location of its representation in geographic software, that is, a conceptualization reflecting much of digital humanitarian research outside disciplinary geography. Goodchild and Glennon implicitly posit that VGI comprises observable, material phenomena having a Cartesian (latitudinal and longitudinal) positionality, and *accuracy* is the measure of difference between this position and its representation. This is at odds with other potential understandings of knowledge representations, for instance, the notion that knowledges can be interpersonal, emotional, performative, tactical, and associated with individual and collective memory—in a word, non-Cartesian.[23] Additionally, while digital humanitarianism introduces new big data information flows to emergency management, these data are couched within relations, behaviors, and norms.[24] All of these attributes affect the data and the practices of data producers and emergency managers. In other words, a limited conception of big data digital humanitarianism would treat it as merely data.

Some researchers claim that big data, regardless of its accuracy, influences emergency managers' operations and decision-making practices. According to these views, mapping activities collect knowledge of *where* emergencies are occurring and presumably *where* aid and support should thus be distributed.[25] For instance, Roche et al. describe the ways new information flow structures embedded in the geoweb have begun shaping emergency management practices, arguing that the geoweb entails "key tools for crisis management and communication by all stakeholders: local authorities, emergency respondents, NGOs and the general public."[26] For Roche et al. digital humanitarian technologies primarily contribute new information sources and communication channels, which potentially shape the ways in which the emergency management sector responds to crises. Related research into formal-sector integration of digital humanitarian technologies has made similar claims, tacitly accepting that the *presence* of digital humanitarian technologies means the formal sector is indeed *engaging* them in some way.[27]

Recently some researchers have begun proposing theoretical frameworks to explain the social and political implications of digital humanitarianism. Burns has argued that the digital humanitarian case suggests big data should be conceptualized as a new set of practices, epistemologies, and socio-political relations.[28] This stands in contrast with much extant research on big data in digital humanitarianism, which often in its framings disproportionately characterizes the integration as democratizing, empowering, and efficient. Instead this approach highlights the modalities, variegations, and spatial situatedness of big data digital humanitarianism. Further, it indicates many potential further linkages with critical geographic theory, in particular from critical, participatory, and feminist GIS. Bittner et al. suggest combining actor-network theory with critical attention to hegemonic power relations to analyze the social nature of digital humanitarian technologies.[29] They make the important argument that the politics of digital humanitarian technology resides not just in the visual representation of data but also in software code, the positionality of digital humanitarians themselves, and the technological artifacts produced, such as platforms and application program interfaces (APIs). This point is likewise taken up by Burns, who demonstrates ways in which knowledge is politicized, contested, and represented through "moments of closure" in digital humanitarianism.[30] Borrowing from feminist theories of technology, Burns argues that digital humanitarian technologies embody social and political relations that become temporarily solidified in technological artifacts.

The rest of this chapter develops these linkages. In the section that follows, I draw on the research discussed above to suggest ways the empirical phenomenon of big data digital humanitarianism may contribute to broad areas of interest in geoweb research. I show that geoweb research can be enriched and challenged by the particular case of digital humanitarianism, and geoweb scholars could use this case to refine theories of the geoweb. Within literature on the geoweb, I focus primarily on research into its social, political, and economic implications, as this area is imperative for the context of the humanitarian enterprise.

Digital Humanitarianism and the Geoweb

In this section I develop four linkages between big data digital humanitarianism and geoweb research. Above I argued that scholars have not yet fully elucidated the ways these topics manifest in digital humanitarianism, yet I posit that exploring these topics would extend and help refine theorizations of the geoweb. Implicit in this argument is the assumption that geographers are uniquely well positioned to build a theorization of digital humanitarianism.[31] I specifically discuss digital humanitarianism's spatialities, political economy, and knowledge politics.

The Spatialities of Big Data Digital Humanitarianism

Much like humanitarianism writ large, digital humanitarianism involves (re)producing spatial relationships between those who *need* help and those who *deliver* it.[32] It is particularly imperative to examine digital humanitarianism in this regard, since the spatial relationship is extended to include those contributing from afar. Collaborators on humanitarian projects no longer need be near the site of the crisis or in the management offices and headquarters; individuals may instead produce, process, and represent data while being located anywhere with an Internet connection. This is accomplished through the mobilization of spaces specific to digital humanitarianism, including its software platforms, lines of code, data models, and data representations. These are distinctive material spaces, both in the sense that their lines of code affect absolute geographies through the distribution of aid and resources and in the sense that they cannot function independent of the fiber optic cables, servers, and real bodies that drive them.[33]

The potential for dispersed digital humanitarianism, however, raises empirical questions regarding who *actually* contributes to these efforts. If big data digital humanitarianism reflects the unevenness of data production observable in other user-generated information efforts, such as Wikipedia, more research will be needed to address the implications of this unevenness.[34] At least one prominent digital humanitarian tool, Ushahidi, originated in the Global South, but its origin says little about current usage, and to date research has not examined the geographic origin of current digital humanitarian contributors.[35] This could be of particular concern if, for instance, remote contributors have little understanding of the roots of the crisis at hand, have little stake in the "success" of the intervention, or have reflected sparsely on how their labor contributes.

Regardless of how geographically dispersed digital humanitarians are, several theoretical questions emerge regarding these spatialities. First, what can be known by those who are remote from the site of the intervention? What does that mean for the types of needs and knowledges that are mapped? In humanitarian crises there must be a relation between distance and the knowledge that one can have of that crisis. This question seeks the epistemic limit of big data digital humanitarianism. Second, what conceptual shifts regarding humanitarianism occur when geographically dispersed laypeople—those with no expert knowledge of how humanitarianism operates—contribute to humanitarian interventions? This may be especially pertinent when big data digital humanitarian projects occur outside sanctioned contexts, as was the case with Mission 4636 in the 2010 Haiti earthquake response, and in ongoing CrowdMap platforms established in the wake of disasters and crises around the globe. Third, what kinds of assumptions about who *has* resources and who is *able to deliver* them motivate digital humanitarian activities? In other words, what must individuals assume about their spatial relationship to those they help, prior to contributing to digital humanitarianism?

These questions resonate with emerging geoweb research exploring the unevenness of user-generated data production. Geoweb research has shown that big data is produced by—and about—the Global North; such data usually reflect and reproduce extant gender relations, as well as influence geographies of consumption.[36] Exploring the spatialities of digital humanitarianism answers Crampton et al.'s call to explore spatialities "beyond the proximate" to understand how humanitarian big data production occurs across diverse spatial scales and relational distances.[37]

The Political Economy of Big Data Digital Humanitarianism

Big data digital humanitarianism is emerging alongside significant political-economic shifts that remain to be explored. These shifts are largely both the result of as well as harbingers of two parallel processes. First, big data digital humanitarianism is often offered as the innovation that can allow the formal humanitarian sector to operate in the context of increased pressures to work more efficiently and effectively.[38] Big data has become prominent within humanitarianism because proponents suggest it will allow humanitarians to fulfill this need, and big data has thus emerged here as the result of political-economic processes within humanitarianism. Second, the private sector has come to play a prominent role in big data digital humanitarianism, as private firms develop many of the technologies needed to produce and effectively leverage big data.[39] Examples of private companies involved in such big data processes include Twitter and Instagram through alert systems, Google Crisis Response in its development of crisis mapping platforms, and Esri in developing emergency management software incorporating digital humanitarian data. Eric Rasmussen, a prominent actant in the digital humanitarian community and CEO of the private business Infinitum Humanitarian Systems, said at a digital humanitarian workshop in 2012, "In these discussions, please let the private sector take care of this. We will address this problem for you, we will take the research, we will commercialize it, and we'll sell it back to you for cheap."[40]

These political-economic processes often go hand-in-hand with a reconceptualization of data vis-à-vis big data. One of the clearest examples comes from Robert Kirkpatrick, director of the UN's Global Pulse program, which seeks to harvest big data for development and humanitarianism. Kirkpatrick has offered a new conceptualization of big data *as aid*.[41] According to this conceptualization, data do not merely capture information about a crisis but also carry material value. In this sense, needs of crisis victims can be met when humanitarian actors circulate big data quickly and efficiently enough.[42] Through this reconceptualization big data becomes a circulating unit of value similar to other commodities, one that can and should be delivered in zones of humanitarian crisis. This therefore constitutes a new locus of political-economic relations and processes that bears critical exploration.

Geoweb scholars have recently begun elucidating the political-economic relations in which the geoweb should be situated. In particular Leszczynski has argued that the geoweb should be understood as signaling both the state stepping back from mapping roles and responsibilities formerly within its purview and as the state stepping forward, in the form of new regulations, frameworks, and markets for the private sector.[43] According to Leszczynski, the geoweb has been made possible by the state handing over its duties to individuals and private sector companies. Much remains to be explored in this area of geoweb research, and big data digital humanitarianism provides a unique case due to its potentially stark consequences. For example, Internet researchers have theorized digital labor in relation to generating surplus value.[44] However, questions remain regarding the reasons private companies are funding many digital humanitarian activities while simultaneously relying on volunteered labor. For what purposes is the private sector entering humanitarianism via digital humanitarianism? What impacts do remote volunteer projects have on local community economies, where in many cases local people are in need of paid work? The case of big data digital humanitarianism raises several questions that could contribute to geoweb scholars' efforts to understand the political economy of the geoweb.

The Knowledge Politics of Big Data Digital Humanitarianism

As in the social sciences more broadly, within digital humanitarianism big data signals new forms of contestation around knowledge legitimacy and differing amounts of weight given to various kinds of claims.[45] In digital humanitarianism these "knowledge politics" take the form of deliberations over what kinds of knowledges to include, how to include them, and the terms on which they should be included.[46] These knowledge politics in digital humanitarianism have implications for how the technologies are engaged by the formal sector and therefore how shifts in aid distribution are to take place. In other words the ways in which digital humanitarian technologies collect, process, and represent big data have implications for how material needs are met.[47]

Much of big data digital humanitarian research has assumed big data contains a degree of representativeness, even if it is simply representative of the small sample of the population producing social media data.[48] Most humanitarian agencies recognize that the knowledges represented in big data are not representative of the population at large.[49] Nevertheless these organizations consider data to be representative of the population that produces it. This view contrasts with recent scholarship showing the *performative* nature of knowledge produced through social media; much recent work has argued that social media data are produced as a way of presenting a desirable image of oneself.[50] Big data within digital humanitarianism could thus be seen not as an objective and situated recording of a reality but rather as a mediated presentation of the image one wants others to have of them. The difference here is between seeing big data as both a source of information to guide the dispersal of resources on the one hand and as a murky lens through which to view a limited number of perspectives of a crisis.

Research into the knowledge politics of the geoweb has shown this to be a fruitful area. While such research has shed light on the uneven representation and contribution of knowledges encoded by data, less is known about the particular ways this has factored into knowledge of the world. Current geoweb research is addressing this question, and digital humanitarianism is an imperative case to consider, as it mobilizes relations that rely explicitly on knowledge of other places. In humanitarian contexts it may be empowering and may assist in recovery and future mitigation efforts for representational technologies to incorporate local knowledges and ways of knowing, rather than to have a structure of knowledge (e.g., prefabricated databases, Cartesian representations) foisted upon local communities. Big data in the form of social media may contribute to this goal, insofar as it allows a degree of flexibility of expression not seen in previous data forms. In this way digital humanitarianism presents unique challenges and ethical questions to knowledge politics in geoweb research. Specific questions remain. Who can—and who *does*—represent places in digital humanitarian contexts? What kinds of marginalizations arise because of these asymmetries? How is knowledge about individual crises, as well as crisis writ large, shaped by digital humanitarian technologies? What kinds of struggles are undertaken by "victims" of humanitarian crises in order to make their knowledges visible—or in some cases to keep their knowledges *in*visible?[51]

Conclusion

In this chapter I have argued that geographers studying the geoweb have not sufficiently considered the case of digital humanitarianism and that doing so would lend unique and productive insights. I have reflected on potential contributions, focusing on the social and political implications of digital humanitarianism. I looked specifically at current theorizations of the geoweb's spatialities, political economy, and knowledge politics, drawing lines of productive resonance with digital humanitarianism.

Further research is needed at the juncture of the geoweb and digital humanitarianism to expand geographers' understanding of the shifting socio-technical practices observed in the world today. Such research would contribute deeper understanding of digital humanitarianism but would also nuance current conceptualizations of the geoweb. As an umbrella term for shifting data and socio-technical phenomena, the "geoweb" currently does not adequately account for the range of contexts in which new data and socio-technical practices occur. Digital humanitarianism enrolls unique institutional, social, and political-economic relations that are distinct from—and therefore stand to refine—what is typically considered the geoweb.

Additionally, greater understanding of digital humanitarianism is imperative due to the significant impact that humanitarianism has on current global social and political relations. The last several decades have witnessed the emergence of a general sense of cosmopolitan responsibility, with appeals to humanitarian and moral sentiments becoming the most likely to generate support for humanitarian intervention.[52] "Humanitarian reason" is thus becoming an increasingly powerful force in the world, and more research is needed to understand the impacts incurred by the incorporation of big data and digital humanitarian technologies.[53] Geoweb scholars are well positioned to address these questions.

Notes

1. Burns, "Rethinking Big Data"; Crawford and Finn, "Limits of Crisis Data."
2. One documented use of this process is found in Standby Task Force, *Nepal Earthquake 2015.*

3. UN OCHA, *Humanitarianism in the Network Age*; Meier, "What Is Big (Crisis) Data?"; Taylor and Schroeder, "Is Bigger Better?"

4. Scharl and Tochtermann, *Geospatial Web*; Leszczynski and Wilson, "Theorizing the Geoweb"; Howe, "Rise of Crowdsourcing"; Ziemke, "Crisis Mapping"; Mayer-Schönberger and Cukier, *Big Data*; Olafsson, *Humanitarian Response*.

5. Crawford and Finn, "Limits of Crisis Data"; Dalton and Thatcher, "Critical Data Studies"; Sandvik et al., "Humanitarian Technology."

6. Goodchild and Glennon, "Crowdsourcing Geographic Information"; Haworth and Bruce, "Review of Volunteered Geographic Information"; Li and Goodchild, "Role of Social Networks in EM"; Hughes and Palen, "Evolving Role"; Starbird and Palen, "Working and Sustaining"; Al-Akkad and Raffelsberger, "How Do I Get This App?"; Hiltz, Kushma, and Plotnick, "Use of Social Media"; Van Gorp, "Integration of Volunteer and Technical Communities."

7. Burns and Shanley, *Connecting Grassroots to Government*; Hiltz, Kushma, and Plotnick, "Use of Social Media"; Shanley et al., "Tweeting Up a Storm."

8. Meier, "Debating the Value of Tweets"; Meier, "What Was Novel?"

9. Gupta and Kumaraguru, "Credibility Ranking of Tweets"; Vieweg, "Situational Awareness"; Castillo, Mendoza, and Poblete, "Predicting Information Credibility."

10. Cohen, "Sandy Marked Shift."

11. Crowley and Chan, *Disaster Relief 2.0*; Burns, "Rethinking Big Data."

12. Palen, Vieweg, and Anderson, "Supporting 'Everyday Analysts'"; Hughes and Palen, "Evolving Role."

13. F. Harvey, "To Volunteer or to Contribute?"; Schroeder, "Big Data"; Liu et al., "How Disaster Information"; Starbird and Palen, "Working and Sustaining."

14. Acar and Muraki, "Twitter for Crisis Communication"; Verity, *OCHA's Lessons Learned*.

15. Corbett and Keller, "Using Community Information."

16. Bott, Gigler, and Young, *Role of Crowdsourcing*; Stottlemyre and Stottlemyre, "Crisis Mapping Intelligence."

17. Chamales and Baker, *Securing Crisis Maps*; Chamales, *Towards Trustworthy Social Media*; Goolsby, *On Cybersecurity, Crowdsourcing*; Meier, "Humanitarians in the Sky"; Sandvik and Lohne, "Rise of the Humanitarian Drone"; Burns and Shanley, *Connecting Grassroots to Government*.

18. Capelo, Chang, and Verity, *Guidance for Collaborating*; Waldman, Verity, and Roberts, *Guidance for Collaborating*; Virtual Social Media Working Group and DHS First Responders Group, *Lessons Learned*.

19. Burns, "Rethinking Big Data."

20. Burns and Shanley, *Connecting Grassroots to Government*; Shanley, "Opportunities and Challenges"; Su, Wardell, and Thorkildsen, *Social Media in Emergency Management*; Woodrow Wilson Center, *Legal and Policy Issues*.

21. Robson, *Responding to Liability*; Reidenberg et al., "Privacy and Missing Persons"; Young et al., *Transforming Earthquake Detection*.

22. Goodchild and Glennon, "Crowdsourcing Geographic Information."

23. Elwood and Mitchell, "Another Politics Is Possible"; Kinsley, "Beyond the Screen"; Kinsley, "Matter of 'Virtual' Geographies"; W. Lin, "Situating Performative Neogeography."

24. Bowker, *Memory Practices*; Boyd and Crawford, "Critical Questions for Big Data"; Dalton and Thatcher, "Critical Data Studies."

25. Zook et al., "Volunteered Geographic Information and Crowdsourcing."

26. Roche et al., "GeoWeb and Crisis Management," 17.

27. Liu and Pale, "New Cartographers"; Liu, Fraustino, and Jin, "How Disaster Information"; Meier, *Digital Humanitarians*.

28. Burns, "Rethinking Big Data."

29. Bittner et al., "Tracing Contingencies."

30. Burns, "Moments of Closure."

31. Similar arguments for geography and big data may be found in Farmer and Pozdnoukhov, "Building Streaming GIScience."

32. Fassin, *Humanitarian Reason*.

33. Kitchin and Dodge, *Code/Space*; Kinsley, "Matter of 'Virtual' Geographies."

34. Suggesting that big data digital humanitarianism reflects the unevenness of data production are M. Graham, Hale, and Stephens, *Geographies of World's Knowledge*; and M. Graham et al., "Uneven Geographies."

35. Okolloh, "Ushahidi or 'Testimony.'"

36. Haklay, "Neogeography and Delusion"; M. Graham et al., "Uneven Geographies"; Stephens, "Gender and Geoweb"; Leszczynski and Elwood, "Feminist Geographies"; M. Graham, "Neogeography and Palimpsests"; Thatcher, "Avoiding the Ghetto."

37. Crampton et al., "Beyond the Geotag."

38. Kirkpatrick, "Data Philanthropy"; Letouzé, "Big Data for Development"; Letouzé, "Big Data from Cellphones"; UN OCHA, *Humanitarianism in the Network Age*.

39. Burns and Shanley, *Connecting Grassroots to Government*.

40. Quoted in Woodrow Wilson Center, *Research Challenges*. To be sure, Rasmussen's statement was paraphrasing another person's words, but it was in affirmation of those words.

41. Kirkpatrick, "Data Philanthropy."
42. Decker, "Big Data for Humanitarian Assistance"; Meier, "What Was Novel?"
43. Leszczynski, "Situating the Geoweb."
44. Fuchs and Sevignani, "What Is Digital Labour?"
45. Elwood and Mitchell, "Another Politics Is Possible."
46. Burns, "Moments of Closure."
47. Burns, "Rethinking Big Data."
48. Letouzé, "Big Data for Development"; Meier, "What Is Big (Crisis) Data?"; UN OCHA, *Humanitarianism in the Network Age.*
49. Meier, "Debating the Value of Tweets."
50. Kinsley, "Beyond the Screen"; W. Lin, "Situating Performative Neogeography."
51. Burns, "Moments of Closure"; Young and Gilmore, "Subaltern Empowerment."
52. Calhoun, "World of Emergencies"; Roy, *Poverty Capital*; Fassin, *Humanitarian Reason.*
53. Calhoun, "World of Emergencies"; Crawford and Finn, "Limits of Crisis Data."

PART 6
Conclusions

12 Rethinking the Geoweb and Big Data

Future Research Directions

Mark Graham

This short chapter is a reflection on future directions that research on the geoweb and big data could take. It is derived from a reflection that the editors of this volume asked me to provide to a session on the geoweb and big data at the 2014 meeting of the Association of American Geographers. Panelists were asked to summarize some of the day's themes and to debate how they speak to future directions in the discipline. My reflections are organized into seven themes.

Defining the Object of Our Inquiry

It is important to point out that different researchers are referring to very different things with the terms "big data" and "geoweb." This is not necessarily a problem, but I hope that this book has helped to clarify what exactly the geoweb is. Words, as signifiers, are able to not just represent but also reproduce and enact change in the social world.[1]

As such, it is crucial to unpack what we mean and what we are trying to convey when we refer to the geoweb. Where are the boundaries between the web and the geoweb? (I am not sure that I clearly see them.) Where are the boundaries between the geoweb and what we might think of as the underlying/offline/material geo that seems to underpin, augment, or inform it? I'm also not sure I clearly see those boundaries, in part because of the ways that place is always transduced: constantly remade and reenacted.[2] The point here is not that we need to necessarily agree on any definitions but rather that we should avoid taking for granted some of the assumptions wrapped into these very powerful terms.

Mixed Methods and Brandolini's Law

Much effort has gone into arguing for the need for more mixed-methods research, for less reliance on big data, and for the nuance and context that only qualitative research can provide.[3] These are sentiments that I fully endorse, but here the so-called fourth law of thermodynamics (i.e., "the amount of energy necessary to refute bullshit is an order of magnitude bigger than the amount required to produce it") can potentially provide some useful guidance.

In other words our time is not necessarily best spent when focused on caricatures that we choose to argue against. How many people are actually saying that big data can answer all facets of all societal questions? The answer is, surprisingly few. Let's then move away from arguing for the need for small data and mixed methods and instead begin channeling some of that energy into creating, carrying out, and enacting those hybrid approaches.

Disciplinary Cross-Pollination

This book illustrates some of the potentials for cross-pollination between geography, computer science, information studies, Internet studies, and other social sciences. As many other disciplines are having their own "spatial turns," there is much that we can (and should) contribute as geographers.

But geographers also need to make sure that we are not reinventing the wheel. For instance, there is a growing amount of work in geography that focuses on crowdsourcing and volunteered geographic information.[4] But much of this work being carried out by geographers unfortunately omits research being done in information studies, psychology, and Internet studies that also is trying to understand motivations for crowdsourcing. More could therefore be done to make our disciplinary boundaries a little more porous and allow some of that work to cross over to geography and geoweb research (and then feed back into it).

Human and Machine Labor

Conversations about big data often seem to neglect the truly massive amount of paid human labor that goes into the filtering, sorting, cleaning, manipulating, and managing of it. Big data is often spoken about as something that pings around between sensors, data sets, machines, and algorithms.

But it is ultimately humans, through digital labor, who are creating much of the content that makes up the geoweb.[5] It is thus crucial to keep focusing on and asking critical questions about the digital sweatshops, the micro workers, the click workers, the gold farmers—those laborers in the background who are keeping our networks chugging along.[6] And I hope that we will begin to see more of this work and to remember that automation is often an illusion. What should we be asking about those millions of workers in the shadows, doing unorganized, low-paid, alienated work and making many of our big data ecosystems function?

Privacy and Transparency

Privacy is one of the most pressing issues of our time, and as geographers we are unwisely ceding much of the debate on it to computer scientists (who tend to be informed on the topic) and to politicians (who don't). Because many privacy implications center on location—its collection, its data structures, its precision, and its accuracy—we can ask what we should be doing and saying and researching as geographers, to draw on our expertise and the strengths of our discipline to make a difference in this new world of always-on tracking and monitoring and the datafication of everything.

At the same time, how do we also make sure that privacy isn't used as an excuse for the wholesale locking away of social data by large companies (meaning that we can't use those data to address the social and human questions that really matter)? There have been many projects that have pointed to the value of open geographic data (e.g., Follow the Things, Fair Tracing, and Wikichains).[7] However, it may be that some of these goals stand in direct competition with desires for increased privacy. Where, therefore, do we stand on the transparency/privacy spectrum? And what should we be doing about it?

Digital Exclusion

One of the most important questions that we can ask as researchers of digital information and networks is, What things, what persons, and what places are being left out? Castells points to this issue when he says that "the cost of exclusion from networks increases faster than do the benefits of inclusion in those same networks."[8] This is an area of work that we tend to do well as geographers (and it is a set of questions that people in other disciplines often seem to miss), but it is only a first step. How can we move beyond it? What can or should we do about digital inequalities and exclusions? If we establish that the digital layers that augment place are inherently uneven, unrepresentative, and imbalanced, what can we do with that knowledge? What *should* we do with that knowledge?

We should also think about the other side of this issue. While there has been a lot of focus on sparse data, or contexts where data might not be able to capture the complexities of any given situation, what about contexts where we have too much data? Some of the chapters in this book guide us through methods for dealing with big data, and we need more of this sort of work. Let's have conversations about cluster computing, graph databases, agent-based models, and other methods for grappling with unmanageable volumes of data. Yes, we always need to remember what those data leave out, but unless we want to abandon the whole big data project we should also be—critically—trying to figure out what those data sets *do* tell us about society and how they help us to answer the big questions that we need to ask.

The Big Questions

Finally, let's keep our eyes on the prize. Let's make sure that we are asking the questions that matter and not being too driven by just what data are available. Let's make sure our research continues to focus on questions about things like inequality, power, voice, control, and human welfare.

And I say *continue* because much of this work is already happening (as is evidenced by many of the chapters in this book).

We also can make sure that we are shaping not just the questions being asked but also the data being collected. This may mean doing things such as always being explicit that there is never any such thing as raw data. Data are always socially and humanly constructed and can therefore participate in the constructing and shaping of data.

Platforms, algorithms, and the people and organizations that control them are also playing an ever more central role in the lives of any Internet user. This is because they are making editorial decisions that shape our lives.[9] Organizations such as Google and Apple and platforms like Wikipedia and Facebook are arbiters of not just what we see and read but of what we know about our world and how we navigate through our world.

Therefore, as the geoweb becomes ever more integral to our lives, it might be instructive to turn to some words spoken by the late Tony Benn, a British Labour Party politician. He famously had a set of five questions he said that we should always ask any powerful person: "What power have you got? Where did you get it from? In whose interests do you exercise it? To whom are you accountable? And how can we get rid of you?"[10]

Many of the platforms and mediators that we rely on for geographic information are for-profit entities that do not necessarily have issues of justice, equality, human rights, and peace at the top of their list of concerns. We could therefore use moments of reflection to dream up and practice radical and democratic alternatives in the geoweb. At the very least we should use Tony Benn's provocations to hold the data intermediaries, systems, platforms, and algorithms accountable for the power that they wield.

Let's continue to ask, "What power have you got? Where did you get it from? In whose interests do you exercise it? To whom are you accountable? And how can we get rid of you?"

Notes

1. I. Parker, *Discourse Dynamics*, 1; Howarth, *Discourse*, 9–12.
2. M. Graham, Zook, and Boulton, "Augmented Reality."

3. Ruppert, "Rethinking Empirical Social Sciences"; Gorman, "Danger of a Big Data Episteme"; M. Graham and Shelton, "Geography and the Future of Big Data."

4. Sui, Elwood, and Goodchild, *Crowdsourcing Geographic Knowledge.*

5. Scholz, *Digital Labor*; Fuchs, *Digital Labour and Karl Marx*; Lehdonvirta and Ernkvist, *Knowledge Map of the Virtual Economy*; M. Graham, Hjorth, and Lehdonvirta, "Digital Labour and Development."

6. There are large and nascent networks of researchers already tackling these issues (see, for instance, the groups formed around the Institute for Distributed Creativity mailing list or the ERC COST network on the dynamics of virtual work). My own work (e.g., M. Graham, Hjorth, and Lehdonvirta, "Digital Labour and Development") also increasingly focuses on questions about who benefits and who doesn't from low-skilled forms of digital labor.

7. Kleine, "Negotiating Partnerships"; M. Graham and Haarstad, "Transparency and Development."

8. Castells, "Network Theory of Power," 774.

9. Kitchin and Dodge, *Code/Space*; M. Graham, Zook, and Boulton, "Augmented Reality."

10. Benn quoted in J. Shaw and Graham, "Informational Right."

BIBLIOGRAPHY

Abraham, J. *Metropolitan Lovers: The Homosexuality of Cities.* Minneapolis: University of Minnesota Press, 2009.

Acar, A., and Y. Muraki. "Twitter for Crisis Communication: Lessons Learned from Japan's Tsunami Disaster." *International Journal of Web Based Communities* 7, no. 3 (2011): 392–402.

Al-Akkad, A., and C. Raffelsberger. "How Do I Get This App? A Discourse on Distributing Mobile Applications despite Disrupted Infrastructure." In *Proceedings of the 11th International ISCRAM Conference,* edited by S. R. Hiltz, M. S. Pfaff, L. Plotnick, and P. C. Shih, 565–69. University Park: Pennsylvania State University, 2014.

Altheide, D. L., and C. J. Schneider. *Qualitative Media Analysis.* 2nd ed. Thousand Oaks CA: SAGE, 2013.

Amoore, L. "Biometric Borders: Governing Mobilities in the War on Terror." *Political Geography* 25 (2006): 336–51.

Anderson, C. "The End of Theory: The Data Deluge Makes the Scientific Method Obsolete." *Wired Magazine* 16, no. 7 (July 2008). http://archive.wired.com/science /discoveries/magazine/16-07/pb_theory.06.23.08.

Anderson, M. "More Americans Using Smartphones for Getting Directions, Streaming TV." Pew Research Center, January 29, 2016. http://www.pewresearch.org /fact-tank/2016/01/29/us-smartphone-use/.

———. "Twitter's Withdrawal of Reliable Share Count API Is a Bold Monetising Move." *TheStack.com,* October 5, 2015. https://thestack.com/cloud/2015/10/05 /twitters-withdrawal-of-reliable-share-count-api-is-a-bold-monetising-move/.

Andrejevic, M. *Infoglut: How Too Much Information Is Changing the Way We Think and Know.* New York: Routledge, 2013.

Andrienko, G., N. Andrienko, D. Keim, A. MacEachren, and S. Wrobel. "Challenging Problems of Geospatial Visual Analytics." *Journal of Visual Languages and Computing* 22, no. 4 (2011): 251–56.

Anthes, G. "Digital Divide Persists, Berkeley Study Shows." *Communications of the ACM* 54, no. 8 (2011): 20.

Apple. "TERMS AND CONDITIONS." 2015. https://www.apple.com/legal/internet
-services/itunes/us/terms.html.

Armstrong, P., and H. Armstrong. "Beyond Numbers: Problems with Quantitative
Data." *Alternate Routes: A Journal of Critical Social Research* 6 (1983): 1–47. http://
www.alternateroutes.ca/index.php/ar/article/view/20273.

Arrhenius, S. "On the Influence of Carbonic Acid in the Air upon the Temperature of
the Ground." *Philosophical Magazine and Journal of Science* 5, no. 41 (1896): 237–76.

Aspray, W., and M. Gunderloy. "Early Computing and Numerical Analysis at the
National Bureau of Standards." *Annals of the History of Computing* 11, no. 1 (1989): 3–12.

Association of American Geographers. *Global Change and Local Places: Estimating,
Understanding, and Reducing Greenhouse Gases.* Cambridge: Cambridge University
Press, 2003.

Bailey, T., and A. Gatrell. *Interactive Spatial Analysis.* London: Longman, 1995.

Baldwin, H. "Big Data as Boogeyman." *Forbes,* October 20, 2014. http://www.forbes
.com/sites/howardbaldwin/2014/10/20/big-data-as-boogeyman/.

Ballantyne, B. "Geomatics and the Law: In Praise of Small Data; Survey Field Notes
in Litigation." *Geomatica* 69, no. 1 (2015): 122–24.

Bamford, J. *The Shadow Factory: The Ultra-Secret NSA from 9/11 to the Eavesdropping
on America.* New York: Anchor Books, 2009.

Baraniuk, R. "More Is Less: Signal Processing and the Data Deluge." *Science* 331, no.
6018 (2011): 717–19. http://science.sciencemag.org/content/331/6018/717.

Barber, K., and K. Harfoush. "Synchronizing Small Data in a Big World." 2014 IEEE Symposium on Computers and Communication, Madeira, Portugal, June 23–26, 2014.

Barbrook, R. *Imaginary Futures: From Thinking Machines to the Global Village.* London:
Pluto Press, 2007.

Bard, K., and S. Shubert. *Encyclopedia of the Archaeology of Ancient Egypt.* London:
Routledge, 1999.

Barnes, T. "Big Data, Little History." *Dialogues in Human Geography* 3, no. 3 (2013):
297–302. doi:10.1177/2043820613514323.

———. "Not Only . . . But Also': Quantitative and Critical Geography." *Professional
Geographer* 61, no. 3 (2009): 292–300.

———. "Retheorizing Economic Geography: From the Quantitative Revolution
to the 'Cultural Turn.'" *Annals of the Association of American Geographers* 91, no. 3
(2001): 546–65.

Barnes, T., and E. Sheppard. "'Nothing Includes Everything': Towards Engaged Pluralism in Anglophone Economic Geography." *Progress in Human Geography* 34,
no. 2 (2009): 193–214.

Barnes, T., and M. Wilson. "Big Data, Social Physics, and Spatial Analysis: The Early Years." *Big Data and Society* 1, no. 1 (2014): 1–14. doi:10.1177/2053951714535365.

Bar-Yam, Y. "The Limits of Phenomenology: From Behaviorism to Drug Testing and Engineering Design." New England Complex Systems Institute (NECSI) Report, August 1, 2013. http://arxiv.org/pdf/1308.3094v1.pdf.

Batty, M. "Big Data, Smart Cities and City Planning." *Dialogues in Human Geography* 3, no. 3 (2013): 274–79.

——. "Editorial." *Environment and Planning B: Planning and Design* 39 (2012): 191–93.

Baxter, J., and J. Eyles. "Evaluating Qualitative Research in Social Geography: Establishing 'Rigour' in Interview Analysis." *Transactions of the Institute of British Geographers* 22, no. 4 (1997): 505–25. http://onlinelibrary.wiley.com/doi/10.1111/j.0020-2754.1997.00505.x/abstract.

Becker, M., R. Alzahabi, and C. Hopwood. "Media Multitasking Is Associated with Symptoms of Depression and Social Anxiety." *Cyberpsychology, Behavior, and Social Networking* 16, no. 2 (2013): 132–35.

Bell, G., T. Hey, and A. Szalay. "Beyond the Data Deluge." *Science* 323, no. 5919 (2009): 1297–98. http://science.sciencemag.org/content/323/5919/1297.

Beller-Simms, N., and H. Ingram. *Decision-Support Experiments and Evaluations Using Seasonal-to-Interannual Forecasts and Observational Data: A Focus on Water Resources; Synthesis and Assessment Product 5.3 Report.* US Climate Change Science Program, 2008.

Benda, D. "Sharing Information on Social Media Invites ID Theft." Scripps Howard News Service, October 14, 2010. http://www.theledger.com/article/20101014/NEWSCHIEF/10145023.

Benjamin, W. *The Arcades Project.* Cambridge MA: Belknap Press of Harvard University Press, 1999.

Berger, D. "Balancing Consumer Privacy with Behavioral Targeting." *Santa Clara Computer and High Technology Law Journal* 27 (2011): 3.

Bernd, J., B. Gordo, J. Choi, B. Morgan, N. Henderson, S. Egelman, D. Garcia, and G. Friedland. "Teaching Privacy: Multimedia Making a Difference." *MultiMedia, IEEE* 22 (2015): 12–19.

Bernholdt, D., S. Bharathi, D. I. Brown, K. Chanchio, M. Chen, A. Chervenak, L. Cinquini, B. Drach, I. Foster, P. Fox, J. Garcia, C. Kesselman, R. Markel, D. E. Middleton, V. Nefedova, L. Pouchard, A. Shoshani, A. Sim, W. G. Strand, and D. Williams. "The Earth System Grid: Supporting the Next Generation of Climate Modeling Research." *Proceedings of the IEEE* 93, no. 3 (2005): 485–95. doi:10.1109/JPROC.2004.842745.

Bettini, C., and D. Riboni. "Privacy Protection in Pervasive Systems: State of the Art and Technical Challenges." *Journal of Pervasive and Mobile Computing* 17, part B (2015): 159–74.

Bittner, C., G. Glasze, and C. Turk. "Tracing Contingencies: Analyzing the Political in Assemblages of Web 2.0 Cartographies." *GeoJournal* 78, no. 6 (2013): 935–48.

Blank, G. "Who Creates Content? Stratification and Content Creation on the Internet." *Information Communication and Society* 16, no. 4 (2013): 590–612. doi:10.1080/1369118x.2013.777758.

Blatt, A. *Health, Science, and Place.* New York: Springer International, 2015.

Bobkowski, P., and J. Smith. "Social Media Divide: Characteristics of Emerging Adults Who Do Not Use Social Network Websites." *Media Culture and Society* 35, no. 6 (2013): 771–81. doi:10.1177/0163443713491517.

Bohli, J., D. Dobre, G. Karame, and W. Li. "PrivLoc: Preventing Location Tracking in Geofencing Services." In *Trust and Trustworthy Computing,* edited by T. Holz and S. Ioannidis, 143–60. New York: Springer International, 2014.

Bonde, A. "Defining Small Data." *Small Data Group,* October 18, 2013. http://smalldatagroup.com/2013/10/18/defining-small-data.

Borsook, P. *Cyberselfish: A Critical Romp through the Terribly Libertarian Culture of High Tech.* New York: Public Affairs, 2000.

Bott, M., B. Gigler, and G. Young. *The Role of Crowdsourcing for Better Governance in Fragile State Contexts,* 2012. http://www.scribd.com/doc/80707030/The-Role-of-Crowdsourcing-for-Better-Governance-in-Fragile-State-Contexts.

Bowker, G. *Memory Practices in the Sciences.* Cambridge MA: MIT Press, 2005.

Bowker, G., and L. Star. *Sorting Things Out: Classification and Its Consequences.* Cambridge MA: MIT Press, 1999.

boyd, d., and K. Crawford. "Critical Questions for Big Data: Provocations for a Cultural, Technological, and Scholarly Phenomenon." *Information, Communication and Society* 15, no. 5: (2012): 662–79. doi:10.1080/1369118x.2012.678878.

———. "Six Provocations for Big Data." Presented at Oxford Internet Institute's A Decade in Internet Time: Symposium on the Dynamics of the Internet and Society, Oxford UK, 2012.

boyd, d., and N. Ellison. "Social Network Sites: Definition, History, and Scholarship." *Journal of Computer-Mediated Communication* 13 (2008): 210–30.

Brandtzæg, P., J. Heim, and A. Karahasanović. "Understanding the New Digital Divide—A Typology of Internet Users in Europe." *International Journal of Human-Computer Studies* 69, no. 3 (2011): 123–38. doi:10.1016/j.ijhcs.2010.11.004.

Braun, J., and T. Gillespie. "Hosting the Public Discourse, Hosting the Public." *Journalism Practice* 5, no. 4 (2011): 383–98.

Brewer, J., and A. Hunter, eds. *Foundations of Multimethod Research: Synthesizing Styles.* Thousand Oaks CA: SAGE, 2006.

Brickner, M., D. Sheffer, Y. Alef, I. Brickner, and A. Sirkis. "Better Decision Making through Representation and Reduction of Uncertainty in C3I Information System." DTIC document, Department of Defense, November 2007. http://oai.dtic.mil /oai/oai?verb=getRecordandmetadataPrefix=htmlandidentifier=ada478785.

Brighenti, A. "New Media and Urban Motilities: A Territoriologic Point of View." *Urban Studies* 49, no. 2 (2012): 399–414. doi:10.1177/0042098011400771.

Brin, S., and L. Page. "The Anatomy of a Large-Scale Hypertextual Web Search Engine." Paper presented at the Seventh International World-Wide Web Conference (www 1998), Brisbane, Australia, April 14–18, 1998. http://ilpubs.stanford .edu:8090/361/.

Broos, A., and K. Roe. "The Digital Divide in the Playstation Generation: Self-Efficacy, Locus of Control and ICT Adoption among Adolescents." *Poetics* 34, no. 4–5 (2006): 306–17. doi:10.1016/j.poetic.2006.05.002.

Brown, C., and R. Wilby. "An Alternate Approach to Assessing Climate Risks." *Eos: Transactions of the American Geophysical Union* 93, no. 41 (2012): 401–12.

Brown, I. "Social Media Surveillance." In *The International Encyclopedia of Digital Communication and Society,* edited by R. Mansell and P. Ang, 1–7. Hoboken NJ: Wiley, 2015.

Brown, R., D. Barram, and L. Irving. *Falling through the Net: A Survey of the "Have Nots" in Rural and Urban America.* Washington DC: National Telecommunications and Information Administration, U.S. Department of Commerce, 1995. http:// www.ntia.doc.gov/ntiahome/fallingthru.html.

Brynjolfsson, E., and A. McAfee. "The Big Data Boom Is the Innovation Story of Our Time." *The Atlantic,* November 2011. http://www.theatlantic.com/business/archive /2011/11/the-big-data-boom-is-the-innovation-story-of-our-time/248215/.

Buhl, H., M. Röglinger, F. Moser, and J. Heidemann. "Big Data: A Fashionable Topic with(out) Sustainable Relevance for Research and Practice?" *Business and Information Systems Engineering* 5, no. 2 (2013): 65–69.

Buneman, P., S. Khanna, and W. Tan. "Data Provenance: Some Basic Issues." In *FST TCS 2000: Foundations of Software Technology and Theoretical Computer Science* SE 6, edited by S. Kapoor and S. Prasad, 87–93. Berlin: Springer. doi:10.1007/3 -540-44450-5_6.

Burdick, A., K. Drucker, P. Lunenfeld, T. Presner, and J. Schnapp. *Digital_Humanities.* Cambridge MA: MIT Press, 2012.

Burns, R. "Moments of Closure in the Knowledge Politics of Digital Humanitarianism." *Geoforum* 53 (May 2014): 51–62.

————. "Rethinking Big Data in Digital Humanitarianism: Practices, Epistemologies, and Social Relations." *GeoJournal* 80, no. 4 (2015): 477–90. http://link.springer.com/article/10.1007/s10708-014-9599-x.

Burns, R., and L. Shanley. *Connecting Grassroots to Government for Disaster Management: Workshop Report.* Washington DC: Woodrow Wilson International Center for Scholars, 2013. http://www.scribd.com/doc/165813847/Connecting-Grassroots-to-Government-for-Disaster-Management-Workshop-Summary.

Burns, R., and J. Thatcher. "Guest Editorial: What's So Big about Big Data? Finding the Spaces and Perils of Big Data." *GeoJournal* 80, no. 4 (2014): 445–48.

Burrows, R., and M. Savage. "After the Crisis? Big Data and the Methodological Challenges of Empirical Sociology." *Big Data and Society* 1, no. 1 (2014): 1–6.

Buytendijk, F. *Hype Cycle for Big Data.* Stamford CT: Gartner, August 4, 2014. https://www.gartner.com/doc/2814517/hype-cycle-big-data-.

Calhoun, C. "A World of Emergencies: Fear, Intervention, and the Limits of Cosmopolitan Order." *Canadian Review of Sociology and Anthropology* 41, no. 4 (2004): 373–95.

Campbell, S., and N. Kwak. "Mobile Communication and Civic Life: Linking Patterns of Use to Civic and Political Engagement." *Journal of Communication* 60, no. 3 (2010): 536–55.

Capelo, L., N. Chang, and A. Verity. *Guidance for Collaborating with Volunteer and Technical Communities.* Geneva: Digital Humanitarian Network, 2012. http://www.scribd.com/doc/101420316/Guidance-for-Collaborating-With-v-amp-TCs.

Castells, M. "A Network Theory of Power." *International Journal of Communications* 5 (2011): 773–87.

Casti, J. *Would-Be Worlds: How Simulation Is Changing the Frontiers of Science.* New York: Wiley, 1997.

Castillo, C., M. Mendoza, and B. Poblete. "Predicting Information Credibility in Time-Sensitive Social Media." *Internet Research* 23, no. 5 (2013): 560–88.

Cavoukian, A. *Privacy by Design: A Primer.* 2009. http://www.privacybydesign.ca/content/uploads/2013/10/pbd-primer.pdf.

CBS News. "Social Media Related Crimes." *Crimesider*, CBSNews.com, 2015. http://www.cbsnews.com/pictures/social-media-related-crimes/.

Ceruzzi, P. *A History of Modern Computing.* 2nd ed. Cambridge MA: MIT Press, 2003.

Chamales, G. *Towards Trustworthy Social Media and Crowdsourcing.* Washington DC: Woodrow Wilson International Center for Scholars, 2013. http://www.scribd.com/doc/138508756/Towards-Trustworthy-Social-Media-and-Crowdsourcing.

Chamales, G., and R. Baker. *Securing Crisis Maps in Conflict Zones*. Seattle WA: Rogue Genius, 2011. http://roguegenius.com/wp-content/uploads/2011/11/Securing -Crisis-Maps-in-Conflict-Zones-Chamales_BakerIEEE-Formatted.pdf.

Chambers, R. "Participatory Mapping and Geographic Information Systems: Whose Map? Who Is Empowered and Who Disempowered? Who Gains and Who Loses?" *Electronic Journal of Information Systems in Developing Countries* 25, no. 2 (2006): 1–11.

Chapman, K. "Community Mapping for Exposure in Indonesia." *HOT Recent Updates* (blog), Humanitarian OpenStreetMap Team, 2012. http://hot.openstreetmap .org/updates/2012-07-31_indonesian_project_report_get_it_while_its_hot.

Chen, J., Y. Chen, X. Du, C. Li, J. Lu, S. Zhao, and X. Zhou. "Big Data Challenge: A Data Management Perspective." *Frontiers of Computer Science* 7, no. 2 (2013): 157–64.

Chen, M., S. Mao, Y. Liu. "Big Data: A Survey." *Mobile Networks and Applications* 19, no. 2 (2014): 171–209.

Cheng, Pei-Sze. "Facebook Chat Led to NJ Burglary." *NBCNewYork.com*, March 25, 2012. http://www.nbcnewyork.com/news/local/Facebook-Chat-Burglary-Family -Friends-Park-Ridge-New-Jersey-144139685.html.

Chorley, R., and P. Haggett, eds. *Models in Geography*. London: Methuen, 1967.

Chun, W. *Control and Freedom: Power and Paranoia in the Age of Fiber Optics*. Cambridge MA: MIT Press, 2006.

Cilliers, P. *Complexity and Postmodernism: Understanding Complex Systems*. London: Routledge, 1998.

CIPPIC. *On the Data Trail: How Detailed Information about You Gets into the Hands of Organizations with Whom You Have No Relationship; A Report on the Canadian Data Brokerage Industry*. Canadian Internet Public Policy Interest Clinic, 2006. https://www.cippic.ca/sites/default/files/May1-06/DatabrokerReport.pdf.

———. *Submissions to the House of Commons Standing Committee on Access to Information, Privacy and Ethics*. Canadian Internet Public Policy Interest Clinic, 2002. https://cippic.ca/en/ETHI_hearings_on_Social_Media_Privacy.

CitiBike. "System Data." *CitiBikeNYC.com*. http://www.citibikenyc.com/system-data.

Clarke, R. "Information Technology and Dataveillance." *Communications of the ACM* 31, no. 5 (1988): 498–512.

Cohen, J. *Configuring the Networked Self: Law, Code, and the Play of Everyday Practice*. New Haven CT: Yale University Press, 2012.

Cohen, S. "Sandy Marked a Shift for Social Media Use in Disasters." *Emergency Management*, March 7, 2013. http://www.emergencymgmt.com/disaster/Sandy -Social-Media-Use-in-Disasters.html.

Constine, J. "How Big Is Facebook's Data? 2.5 Billion Pieces of Content and 500+ Terabytes Ingested Every Day." *TechCrunch*, August 22, 2012. http://techcrunch.com/2012/08/22/how-big-is-facebooks-data-2-5-billion-pieces-of-content-and-500-terabytes-ingested-every-day/.

Cope, M. "Coding Qualitative Data." In *Qualitative Research Methods in Human Geography*, edited by Iain Hay, 310–24. New York: Oxford University Press, 2005.

——. "Home-Work Links, Labor Markets, and the Construction of Place in Lawrence, Massachusetts, 1920–1939." *Professional Geographer* 50, no. 1 (1998): 126–40.

Cope, M., and S. Elwood, eds. *Qualitative GIS: A Mixed Methods Approach*. London: SAGE, 2009.

Corbett, J., and P. Keller. "Using Community Information Systems to Communicate Traditional Knowledge Embedded in the Landscape." *Participatory Learning and Action* 54 (2006): 21–27.

Cormen, T. *Algorithms Unlocked*. Cambridge MA: MIT Press, 2013.

Cotten, S., T. Hale, M. Moroney, L. O'Neal, and C. Borch. "Using Affordable Technology to Decrease Digital Inequality." *Information, Communication and Society* 14, no. 4 (2011): 424–44.

Couclelis, H. "The Construction of the Digital City." *Environment and Planning B* 31, no. 1 (2004): 5–20.

Couldry, N., and A. Powell. "Big Data from the Bottom Up." *Big Data and Society* 1, no. 2 (2014). doi:2053951714539277.

Coveney, P., and R. Highfield. *Frontiers of Complexity: The Search for Order in a Complex World*. New York: Ballantine Books, 1995.

Crampton, J. "Collect It All: National Security, Big Data and Governance." *GeoJournal* 80, no. 4 (2014): 519–31. http://link.springer.com/article/10.1007/s10708-014-9598-y.

——. *Mapping: A Critical Introduction to Cartography and GIS*. New York: Wiley, 2010.

Crampton, J., M. Graham, A. Poorthuis, T. Shelton, M. Stephens, M. Wilson, and M. Zook. "Beyond the Geotag: Situating 'Big Data' and Leveraging the Potential of the Geoweb." *Cartography and Geographic Information Science* 40, no. 2 (2013): 130–39. doi:10.1080/15230406.2013.777137.

Crampton, J. W., and J. Krygier. "An Introduction to Critical Cartography." *ACME* 4, no. 1 (2006).

Crang, M., T. Crosbie, and S. Graham. "Technology, Time-Space, and the Remediation of Neighbourhood Life." *Environment and Planning A* 39, no. 10 (2007): 2405–22.

Crawford, K. "The Anxieties of Big Data." *The New Inquiry*, May 30, 2014. http://
thenewinquiry.com/essays/the-anxieties-of-big-data.

———. "The Hidden Biases in Big Data." *Harvard Business Review Blog*, April 1,
2013. https://hbr.org/2013/04/the-hidden-biases-in-big-data/.

Crawford, K., and M. Finn. "The Limits of Crisis Data: Analytical and Ethical Challenges of Using Social and Mobile Data to Understand Disasters." *GeoJournal* 80,
no. 4 (2015): 491–502. http://link.springer.com/article/10.1007/s10708-014
-9597-z.

Creswell, J. *Research Design: Qualitative, Quantitative, and Mixed Methods Approaches.*
2nd ed. London: SAGE, 2003.

Crogan, P. *Gameplay Mode: War, Simulation, and Technoculture.* Minneapolis: University of Minnesota Press, 2011.

Crowley, J., and J. Chan. *Disaster Relief 2.0: The Future of Information Sharing in
Humanitarian Emergencies.* Washington DC: UN Foundation; and Berkshire UK:
Vodafone Foundation Technology Partnership, 2011. http://www.globalproblems
-globalsolutions-files.org/gpgs_files/pdf/2011/DisasterResponse.pdf.

Crutcher, M., and M. Zook. "Placemarks and Waterlines: Racialized Cyberscapes
in Post-Katrina Google Earth." *Geoforum* 40, no. 4 (2009): 523–34.

daCosta, F. *Rethinking the Internet of Things: A Scalable Approach to Connecting Everything.* New York: Springer, 2013.

Dalton, C., L. Taylor, and J. Thatcher. "Critical Data Studies: A Dialog on Data and
Space." *Big Data and Society* (forthcoming). Available at SSRN, http://ssrn.com
/abstract=2761166.

Dalton, C., and J. Thatcher. "What Does a Critical Data Studies Look Like, and
Why Do We Care? Seven Points for a Critical Approach to 'Big Data.'" *Society
and Space* (blog), May 12, 2014. http://societyandspace.org/2014/05/12/what
-does-a-critical-data-studies-look-like-and-why-do-we-care-craig-dalton-and-jim
-thatcher/.

Dalton, C. M., and J. Thatcher. "Inflated Granularity: Spatial 'Big Data' and Geodemographics." *Big Data and Society* 2, no. 2 (2015). doi:2053951715601144.

Davenport, T., P. Barth, and R. Bean. "How 'Big Data' Is Different." *MIT Sloan Management Review*, July 30, 2012. http://sloanreview.mit.edu/article/how-big-data
-is-different/.

Davies, T., and M. Frank. "'There's No Such Thing as Raw Data': Exploring the Socio-
Technical Life of a Government Dataset." In *Proceedings of the 5th Annual ACM
Web Science Conference, WebSci 13*, edited by H. Davis, H. Halpin, A. Pentland,

M. Bernstein, L. Adamic, H. Alani, A. Monnin, and R. Rogers, 75–78. New York: ACM, 2013. http://doi.acm.org/10.1145/2464464.2464472.

Decker, K. "Big Data for Humanitarian Assistance and Disaster Response." MS thesis, San Diego State University, 2013.

Debatin, B., J. Lovejoy, A. Horn, and B. Hughes. "Facebook and Online Privacy: Attitudes, Behaviors, and Unintended Consequences." *Journal of Computer-Mediated Communication* 15, no. 1 (2009): 83–108.

Deitrick, S., and R. Edsall. 2006. "The Influence of Uncertainty Visualization on Decision Making: An Empirical Evaluation." In *Progress in Spatial Data Handling*, edited by A. Riedl, W. Kainz, and G. Elmes, 719–38. New York: Springer, 2006.

DeLanda, M. *War in the Age of Intelligent Machines.* New York: Swerve Editions, 1991.

DeLyser, D., and D. Sui. "Crossing the Qualitative-Quantitative Chasm II: Inventive Approaches to Big Data, Mobile Methods, and Rhythmanalysis." *Progress in Human Geography* 37, no. 2 (2013): 293–305.

Denzin, N., and Y. Lincoln, eds. *Strategies of Qualitative Inquiry.* 3rd ed. London: SAGE, 2008.

de Oliveira, M., C. Campelo, C. Baptista, and M. Bertolotto. "Leveraging VGI for Gazetteer Enrichment: A Case Study for Geoparsing Twitter Messages." In *Web and Wireless Geographical Information Systems*, 20–36. New York: Springer International, 2015.

Diebold, Francis. "On the Origin(s) and Development of the Term 'Big Data.'" Penn Institute for Economic Research, Department of Economics, University of Pennsylvania, 2012. http://economics.sas.upenn.edu/sites/economics.sas.upenn.edu/files/12-037.pdf.

DiMaggio, P., E. Hargittai, C. Celeste, and S. Shafer. "Digital Inequality: From Unequal Access to Differentiated Use." In *Social Inequality*, edited by K. Neckerman, 355–400. New York: Russell Sage Foundation, 2004.

Dobson, J., and P. Fisher. "Geoslavery." *Technology and Society Magazine IEEE* 22, no. 1 (2003): 47–52.

Dodge, M., and R. Kitchin. "Codes of Life: Identification Codes and the Machine-Readable World." *Environment and Planning D: Society and Space* 23, no. 6 (2005): 851–81.

———. "Crowdsourced Cartography: Mapping Experience and Knowledge." *Environment and Planning A* 45, no. 1 (2013): 19–36.

———. "Flying through Code/Space: The Real Virtuality of Air Travel." *Environment and Planning A* 36 (2004): 195–211.

———. "Software, Objects, and Home Space." *Environment and Planning A* 41, no. 6 (2009): 1344–65. doi:10.1068/a4138.

Dodge, M., R. Kitchin, and M. Zook. "How Does Software Make Space? Exploring Some Geographical Dimensions of Pervasive Computing and Software Studies." *Environment and Planning A* 41, no. 6 (2009): 1283–93.

Donath, J. *The Social Machine: Designs for Living Online.* Cambridge MA: MIT Press, 2014.

Dourish, P., and M. Mazmanian. "Media as Material: Information Representations as Material Foundations for Organizational Practice." In *How Matter Matters: Objects, Artifacts, and Materiality in Organization Studies,* edited by P. R. Carlisle, 92–118. Oxford: Oxford University Press, 2013.

Duhigg, C. "What Does Your Credit-Card Company Know about You?" *New York Times,* May 17, 2009. http://www.nytimes.com/2009/05/17/magazine/17credit-t.html.

Dunwoody, S. "The Challenge of Trying to Make a Difference Using Media Messages." In *Creating a Climate for Change: Communicating Climate Change and Facilitating Social Change,* edited by S. Moser and L. Dilling, 89–104. Cambridge: Cambridge University Press, 2007.

Dupaquier, M., and J. Dupaquier. *Histoire de la démographie.* Paris: Librairie académique Perrin, 1985.

Dutcher, J. "What Is Big Data?" *datascience@berkeley,* September 3, 2014. http://datascience.berkeley.edu/what-is-big-data/.

Dyson, G. *Turing's Cathedral: The Origins of the Digital Universe.* New York: Pantheon, 2012.

Earle, P., D. Bowden, and M. Guy. "Twitter Earthquake Detection: Earthquake Monitoring in a Social World." *Annals of Geophysics* 54, no. 6 (2011): 708–15.

Eckert, J. "Tropes 2.0: Strategic Mobilizations of the Geoweb." MA thesis, University of Washington, 2010.

Economist. "The Data Deluge: Businesses, Governments and Society Are Only Starting to Tap Its Vast Potential." February 25, 2010. http://www.economist.com/node/15579717.

Elwood, S. "Geographic Information Science: Emerging Research on the Societal Implications of the Geospatial Web" (part I). *Progress in Human Geography* 34, no. 3 (2010): 349–57. doi:10.1177/0309132509340711.

———. "Geographic Information Science: Visualization, Visual Methods, and the Geoweb" (part II). *Progress in Human Geography* 35, no. 3 (2011): 401–8.

———. "Volunteered Geographic Information: Key Questions, Concepts and Methods to Guide Emerging Research and Practice." *GeoJournal* 72, no. 3 (2008): 133–35.

Elwood, S., and A. Leszczynski. "New Spatial Media, New Knowledge Politics." *Transactions of the Institute of British Geographers* 38, no. 4 (2013): 544–59.

———. "Privacy, Reconsidered: New Representations, Data Practices, and the Geoweb." *Geoforum* 42 (2011): 6–15.

Elwood, S., and M. Cope. "Introduction: Qualitative GIS; Forging Mixed Methods through Representations, Analytical Innovations, and Conceptual Engagements." In *Qualitative GIS: A Mixed Methods Approach*, edited by M. Cope and S. Elwood, 1–12. London: SAGE.

Elwood, S., M. Goodchild, and D. Sui. "Prospects for VGI Research and the Emerging Fourth Paradigm." In *Crowdsourcing Geographic Knowledge: Volunteered Geographic Information (VGI) in Theory and Practice*, edited by D. Sui, S. Elwood, and M. Goodchild, 361–75. London: Springer, 2012.

Elwood, S., and K. Mitchell. "Another Politics Is Possible: Neogeographies, Visual Spatial Tactics, and Political Formation." *Cartographica* 48, no. 4 (2013): 275–92.

England, K. "Getting Personal: Reflexivity, Positionality, and Feminist Research." *Professional Geographer* 46, no. 1 (1994): 80–89.

Esteve, A., and M. Sobek. "Challenges and Methods of International Census Harmonization." *Historical Methods* 36, no. 2 (2003): 66–79.

Estrin, D. "Small Data, Where N = Me." *Communications of the ACM* 57, no. 4 (2014): 32–34.

Eubanks, V. *Digital Dead End: Fighting for Social Justice in the Information Age*. Cambridge MA: MIT Press, 2011.

Evans, M., D. Oliver, X. Zhou, and S. Shekhar. "Spatial Big Data: Case Studies on Volume, Velocity, and Variety." In *Big Data: Techniques and Technologies in Geoinformatics*, edited by H. Karimi, 149–76. Boca Raton FL: CRC Press, 2014.

Executive Office of the President. *Big Data: Seizing Opportunities, Preserving Values*. Washington: The White House, 2014. https://obamawhitehouse.archives.gov/sites/default/files/docs/big_data_privacy_report_may_1_2014.pdf.

Farmer, C., and A. Pozdnoukhov. "Building Streaming GIScience from Context, Theory, and Intelligence." In *Proceedings of the Workshop on GIScience in the Big Data Age*. Columbus OH, 2012. http://ncg.nuim.ie/content/staff/staff/downloads/apozdnoukhov/giscience2012.pdf.

Fassin, D. *Humanitarian Reason: A Moral History of the Present*. Berkeley: University of California Press, 2012.

Federal Communications Commission (FCC). *Deployment of Advanced Telecommunications Capability: Second Report*. Washington DC: FCC, 2000.

Feenberg, A. *Critical Theory of Technology*. New York: Oxford University Press, 1991.

Fekete, E. "Consumption and the Urban Hierarchy in the Southeastern United States." *Southeastern Geographer* 54, no. 3 (2014): 249–69.

———. "Race and (Online) Sites of Consumption." *Geographical Review* 105, no. 4 (2015): 472–91.

Fidelman, M. "These Smart, Social Apps Bring Big Data Down to Size." *Forbes*, October 30, 2012. http://www.forbes.com/sites/markfidelman/2012/10/30/these-smart-social-apps-bring-big-data-down-to-size/.

Fiebrich, C. "History of Surface Weather Observations in the United States." *Earth-Science Reviews* 93, no. 3 (2009): 77–84.

Fischetti, M. "The Web Turns 20: Linked Data Gives People Power." *Scientific American*, November 23, 2010. http://www.scientificamerican.com/article/berners-lee-linked-data/.

Flood, J. *The Fires: How a Computer Formula, Big Ideas, and the Best of Intentions Burned Down New York City—and Determined the Future of Cities*. New York: Riverhead Books, 1991.

Floridi, L. "Big Data and Their Epistemological Challenge." *Philosophy and Technology* 25, no. 4 (2012): 435–37.

———. "Data." In *International Encyclopedia of the Social Sciences*, edited by W. Darity. 2nd ed. Detroit: Macmillan, 2008. http://www.philosophyofinformation.net/publications/pdf/data.pdf.

———. *Information: A Very Short Guide*. Oxford: Oxford University Press, 2010.

Foresman, T. *The History of Geographic Information Systems: Perspectives from the Pioneers*. Upper Saddle River NJ: Prentice Hall, 1998.

Foucault, M. "The Confession of the Flesh." In *Power/Knowledge: Selected Interviews and Other Writings, 1972–1977*, edited by C. Gordon, 194–228. New York: Pantheon Books, 1980.

———. *The Birth of Biopolitics: Lectures at the Collège de France, 1978–79*. New York: Picador, 2010.

Foursquare. "About." *Foursquare*. http://www.foursquare.com/about. Last accessed May 8, 2017.

Fox, S. and L. Rainie. *The Internet at 25*. Washington DC: Pew Research Center, 2014. http://www.pewinternet.org/2014/02/27/summary-of-findings-3/.

Fox News World. "Brazil Police Bust Kidnappers Who Browsed Social-Networking Sites for Victims." *FoxNews.com*, August 2, 2010. http://www.foxnews.com/world/2010/08/02/brazil-police-bust-kidnappers-browsed-social-networking-sites-victims/.

Frank, T. *The Conquest of Cool: Business Culture, Counterculture, and the Rise of Hip Consumerism.* Chicago: University of Chicago Press, 1998.

Freudenburg, W., R. Gramling, and D. Davidson. "Scientific Certainty Argumentation Methods (SCAMs): Science and the Politics of Doubt." *Sociological Inquiry* 78, no. 1 (2008): 2–38. doi:10.1111/j.1475-682X.2008.00219.x.

Freudenburg, W., and V. Muselli. "Reexamining Climate Change Debates: Scientific Disagreement or Scientific Certainty Argumentation Methods (SCAMs)?" *American Behavioral Scientist* 57, no. 6 (2013): 777–95. doi:10.1177/0002764212458274.

Frické, M. "Big Data and Its Epistemology." *Journal of the Association for Information Science and Technology* 66, no. 4 (2015): 651–61.

Friedland, G. "Privacy Concerns of Sharing Multimedia in Social Networks." In *Proceedings of the 21st ACM International Conference on Multimedia,* 1121–22. Barcelona: Association for Computing Machinery (ACM), 2013.

Friedland, G., and R. Sommer. "Cybercasing the Joint: On the Privacy Implications of Geo-Tagging." In *Proceedings of HotSec'10: USENIX Workshop on Hot Topics in Security,* Article 1–8. Berkeley CA: USENIX Association, 2013.

Frisch, P. "2016: The Year of the Zettabyte." *Smart Business* (blog). http://storageeffect .media.seagate.com/business/2016-the-year-of-the-zettabyte/.

Fuchs, C. *Digital Labour and Karl Marx.* New York: Routledge, 2014.

Fuchs, C., and S. Sevignani. "What Is Digital Labour? What Is Digital Work? What's Their Difference? And Why Do These Questions Matter for Understanding Social Media?" *tripleC* 11, no. 2 (2013): 237–93.

Fugate, C. "Understanding the Power of Social Media as a Communication Tool in the Aftermath of Disasters." *Address to the U.S. Senate Committee on Homeland Security and Government Affairs, ad hoc Subcommittee on Disaster Recovery and Intergovernmental Affairs, May 5, 2011.* Washington DC: US Government Printing Office, 2011.

Gallagher, S. "Staking Out Twitter and Facebook, New Service Lets Police Poke Perps." *ArsTechnica,* November 13, 2013. http://arstechnica.com/information -technology/2013/11/staking-out-twitter-and-facebook-new-service-lets-police -poke-perps.

Gambs, S., M. Killijian, and M. Cortez. "Show Me How You Move and I Will Tell You Who You Are." In *Proceedings of the 3rd ACM SIGSPATIAL International Workshop on Security and Privacy in GIS and LBS,* 34–41. San Jose CA: Association for Computing Machinery (ACM), 2010.

Gantz, J., and D. Reinsel. *Extracting Value from Chaos.* Framingham MA: IDC, 2011. http://idcdocserv.com/1142.

Garvey, E. "*facts* and FACTS'": Abolitionists' Database Innovations." In *"Raw Data" Is an Oxymoron*, edited by L. Gitelman, 89–102. Cambridge MA: MIT Press, 2013.

Gayo-Avello, D., P. Metaxas, and E. Mustafaraj. "Limits of Electoral Predictions using Social Media Data." *Proceedings of the Fifth International AAAI Conference on Weblogs and Social Media, July 17–21, 2011*, 1–8. Menlo Park CA: AAAI Press, 2011.

General Electric. *Datalandia Trailer—GE*. YouTube, July 16, 2013. https://www .youtube.com/watch?v=FJ-DEQKBWZ4.

Geofeedia. *Case Study: Los Angeles County Sheriff's Department*, 2015. http://resources .geofeedia.com/los-angeles-county-sheriffs-department-case-study.

Ghosh, D., and R. Guha. "What Are We 'Tweeting' about Obesity? Mapping Tweets with Topic Modeling and Geographic Information System." *Cartography and Geographic Information Science* 40, no. 2 (2013): 90–102.

Gieseking, J. *Queer New York: Constellating Geographies of Lesbians' and Queer Women's In/Justice in New York City, 1983–2008* (forthcoming).

Gitelman, L., and V. Jackson. "Introduction." In *"Raw Data" Is an Oxymoron*, edited by L. Gitelman, 1–14. Cambridge MA: MIT Press, 2013.

Glaser, B., and A. Strauss. *The Discovery of Grounded Theory: Strategies for Qualitative Research*. New York: Aldine Transaction, 1967.

Gleick, J. *Chaos: Making a New Science*. New York: Viking Penguin, 1987.

Golumbia, D. *The Cultural Logic of Computation*. Cambridge MA: Harvard University Press, 2009.

Goodchild, M. "Citizens as Sensors: The World of Volunteered Geography." *Geo-Journal* 69, no. 4 (2007): 211–21. doi:10.1007/s10708-007-9111-y.

———. "Commentary: Whither VGI?" *GeoJournal* 72, no. 3–4 (2008): 239–44.

———. "The Quality of Big (Geo)Data." *Dialogues in Human Geography* 3, no. 3 (2013): 280–84.

Goodchild, M., and J. Glennon. "Crowdsourcing Geographic Information for Disaster Response: A Research Frontier." *International Journal of Digital Earth* 3, no. 3 (2010): 231–41.

Google. "Good to Know: A Guide to Staying Safe and Secure Online." https://www .google.com/intl/en_us/goodtoknow/.

Goolsby, R. *On Cybersecurity, Crowdsourcing, and Social Cyber-Attack*. Washington DC: Woodrow Wilson International Center for Scholars, 2013. http://www.scribd.com /doc/127219170/On-Cybersecurity-Crowdsourcing-and-Social-Cyber-Attack -Commons-Lab-Policy-Memo-Series-Vol-1.

Gordon, E., and A. de Souza e Silva. "The Urban Dynamics of Net Localities: How Mobile and Location-Aware Technologies Are Transforming Places." In *Mobile*

Technology and Place, edited by R. Wilken and G. Goggins, 89–103. New York: Routledge, 2012.

Gorman, S. "The Danger of a Big Data Episteme and the Need to Evolve Geographic Information Systems." *Dialogues in Human Geography* 3, no. 2 (2013): 285–91.

Goss, J. "'We Know Who You Are and We Know Where You Live': The Instrumental Rationality of Geodemographic Systems." *Economic Geography* 71, no. 2 (1995): 171–98.

Gotham, K. "Urban Space, Restrictive Covenants and the Origins of Racial Residential Segregation in a US City, 1900–50." *International Journal of Urban and Regional Research* 24, no. 3 (2000): 616–33.

Gould, P. "Is *Statistix Inferens* the Geographical Name for a Wild Goose?" *Economic Geography* 46 (1970): 439–48.

Graham, E., S. Henderson, and A. Schloss. "Using Mobile Phones to Engage Citizen Scientists in Research." *Eos: Transactions of the American Geophysical Union* 92, no. 38 (2011): 313–15. doi:10.1029/2011EO380002.

Graham, M. "Big Data and the End of Theory?" *Guardian,* March 9, 2012. http://www.theguardian.com/news/datablog/2012/mar/09/big-data-theory.

———. "Geography/Internet: Ethereal Alternate Dimensions of Cyberspace or Grounded Augmented Realities?" *Geographical Journal* 172, no. 2 (2013): 177–82.

———. "Neogeography and the Palimpsests of Place: Web 2.0 and the Construction of a Virtual Earth." *Tijdschrift voor Economische en Sociale Geografie* 101, no. 4 (2010): 422–36.

Graham, M., and H. Haarstad. "Transparency and Development: Ethical Consumption through Web 2.0 and the Internet of Things." *Information Technologies and International Development* 7 (2011): 1–18.

Graham, M., S. Hale, and M. Stephens. *Geographies of the World's Knowledge.* Edited by C. M. Flick. London: Convoco! Edition, 2011.

Graham, M., I. Hjorth, and V. Lehdonvirta. "Digital Labour and Development: Impacts of Global Digital Labour Platforms and the Gig Economy on Worker Livelihoods." *Transfer: European Review of Labour and Research,* first posted March 16, 2017. doi:10.1177/1024258916687250.

Graham, M., B. Hogan, R. Straumann, and A. Medhat. "Uneven Geographies of User-Generated Information: Patterns of Increasing Informational Poverty." *Annals of the Association of American Geographers* 104, no. 4 (2014):746–64.

Graham, M., and T. Shelton. "Geography and the Future of Big Data, Big Data and the Future of Geography." *Dialogues in Human Geography* 3, no. 3 (2013): 255–61.

Graham, M., and M. Zook. "Visualizing Global Cyberscapes: Mapping User-Generated Placemarks." *Journal of Urban Technology* 18, no. 1 (2011):115–32.

Graham, M., M. Zook, and A. Boulton. "Augmented Reality in Urban Places: Contested Content and the Duplicity of Code." *Transactions of the Institute of British Geographers* 38, no. 3 (2013): 464–79. doi:10.1111/j.1475-5661.2012.00539.x.

Graham, S. "Software-Sorted Geographies." *Progress in Human Geography* 29, no. 5 (2005): 562–80. doi:10.1191/0309132505ph568oa.

Gray, M. *Out in the Country: Youth, Media, and Queer Visibility in Rural America.* New York: New York University Press, 2009.

Green, G., and A. Goetting. *Mobilizing Communities: Asset Building as a Community Development Strategy.* Philadelphia: Temple University Press, 2010.

Greenberg, P. "10 Reasons 2014 will be the Year of Small Data." *ZDNet*, December 2, 2013. http://www.zdnet.com/article/10-reasons-2014-will-be-the-year-of-small-data.

Grimm, V., E. Revilla, U. Berger, F. Jeltsc, W. Mooij, S. Railsback, H-H. Thulke, J. Weiner, T. Wiegand, and D. DeAngelis. "Pattern-Oriented Modeling of Agent-Based Complex Systems: Lessons from Ecology." *Science* 310 (2007): 987–91.

Groeneveld, F., B. Borsboom, and B. van Amstel. "Over-Sharing and Location Awareness." Washington DC: Center for Democracy and Technology, 2010. https://cdt.org/blog/over-sharing-and-location-awareness/.

Gross, T., and B. Blasius. "Adaptive Coevolutionary Networks: A Review." *Journal of the Royal Society Interface* 5, no. 20 (2008): 259–71.

Guerin, C., S. Carter, and C. Aitchison. "Blogging as Community of Practice: Lessons for Academic Development?" *International Journal for Academic Development* (2015): 212–23.

Gupta, A., and P. Kumaraguru. "Credibility Ranking of Tweets during High Impact Events." In *Privacy and Security in Online Social Media.* Lyon, France: ACM, 2012. http://precog.iiitd.edu.in/Publications_files/a2-gupta.pdf.

Guszcza, J., D. Schweidel, and S. Dutta. "The Personalized and the Personal." *Deloitte Review* 14 (2014): 94–109.

Gutierrez, D. "Big Data vs. Small Data—Is There a Difference?" *insideBIGDATA*, March 24, 2014. http://inside-bigdata.com/2014/03/24/big-data-vs-small-data-difference/.

Hacking, I. "Biopower and the Avalanche of Numbers." *Humanities in Society* 5, no. 3–4 (1982): 279–95.

———. "Philosophie et histoire des concepts scientifiques." In "Façonner les gens" (course description), Collège de France, 2001–2. https://www.college-de-france .fr/media/ian-hacking/UPL7997567846150782232_Hacking2001_2002.pdf.

———. "Kinds of People, Moving Targets." British Academy lecture, April 11, 2006, 285–318. http://socio-logic.jp/2013-hacking-sympo/reading/Hacking_2006 _KindsofPeople_OCR.pdf.

———. "Making Up People." In *Reconstructing Individualism: Autonomy, Individuality, and Self in Western Thought*, edited by T. Heller, D. Wellbery, and M. Sosna, 222–36. Stanford CA: Stanford University Press, 1986.

———. "A Tradition of Natural Kinds." *Philosophical Studies: An International Journal for Philosophy in the Analytic Tradition* 61, no. 1–2 (1991): 109–26.

Hahmann, S., R. Purves, and D. Burghardt. "Twitter Location (Sometimes) Matters: Exploring the Relationship between Georeferenced Tweet Content and Nearby Feature Classes." *Journal of Spatial Information Science* 9 (2014): 1–36. doi:10.5311 /JOSIS.2014.9.185.

Haklay, M. "Citizen Science and Volunteered Geographic Information: Overview and Typology of Participation." In *Crowdsourcing Geographic Knowledge: Volunteered Geographic Information (VGI) in Theory and Practice*, edited by D. Sui, S. Elwood, and M. Goodchild, 105–22. London: Springer, 2012.

———. "Geographic Information Science: Tribe, Badge and Sub-Discipline." *Transactions of the Institute of British Geographers* 37, no. 4 (2012): 477–81.

———. "Neogeography and the Delusion of Democratisation." *Environment and Planning A* 45, no. 1 (2013): 55–69.

Haklay, M., A. Singleton, and C. Parker. "Web Mapping 2.0: The Neogeography of the GeoWeb." *Geography Compass* 2, no. 6 (2008): 2011–39. doi:10.1111/j.1749 -8198.2008.00167.x.

Halberstam, J. *In a Queer Time and Place: Transgender Bodies, Subcultural Lives*. New York: New York University Press, 2005.

Hampton, K. "Internet Use and the Concentration of Disadvantage: Glocalization and the Urban Underclass." *American Behavioral Scientist* 53, no. 8 (2010): 1111–32. doi:10.1177/0002764209356244.

Hannah, M. *Dark Territory in the Information Age*. Farnham UK: Ashgate, 2011.

Haraway, D. *A Cyborg Manifesto: Science, Technology, and Socialist-Feminism in the Late Twentieth Century*. London: Routledge, 1991.

———. *Modest_Witness@Second_Millennium: FemaleMan_Meets_OncoMouse; Feminism and Technoscience*. New York: Routledge, 1997.

Harcourt, B. *Against Prediction: Profiling, Policing, and Punishing in an Actuarial Age*. Chicago: University of Chicago Press, 2006.

Harding, S. "After the Neutrality Ideal: Science, Politics, and 'Strong Objectivity.'" *Social Research* 59, no. 3 (1992): 567–87. http://www.jstor.org/stable/pdf/40970706.pdf.

Hardt, M. "How Big Data Is Unfair." *Medium*, September 26, 2014. https://medium.com/@mrtz/how-big-data-is-unfair-9aa544d739de.

Hargittai, E., and A. Hinnant. "Digital Inequality: Differences in Young Adults' Use of the Internet." *Communication Research* 35, no. 5 (2008): 602–21.

Harrower, M. "Representing Uncertainty: Does It Help People Make Better Decisions." In *UCGIS Workshop: Geospatial Visualization and Knowledge Discovery Workshop*, 1–7. Ithaca NY: University Consortium for Geographic Information Science, 2003. http://citeseerx.ist.psu.edu/viewdoc/download?doi=10.1.1.59.2463&rep=rep1&type=pdf.

Hartwick, E. "Geographies of Consumption: A Commodity-Chain Approach." *Environment and Planning D: Society and Space* 16 (1998): 423–37.

Harvey, D. *The Condition of Postmodernity: An Enquiry into the Origins of Cultural Change.* New York: Wiley, 1990.

———. *The Enigma of Capital and the Crises of Capitalism.* Oxford: Oxford University Press, 2010.

Harvey, D., and M. Reed. "Social Science as the Study of Complex Systems." In *Chaos Theory in the Social Sciences: Foundations and Applications*, edited by L. Kiel and E. Elliott, 295–323. Ann Arbor: University of Michigan Press, 1996.

Harvey, F. "To Volunteer or to Contribute Locational Information? Towards Truth in Labeling for Crowdsourced Geographic Information." In *Crowdsourcing Geographic Knowledge: Volunteered Geographic Information (VGI) in Theory and Practice*, edited by D. Sui, S. Elwood, and M. Goodchild, 31–42. London: Springer, 2012.

Hawelka, B., I. Sitko, E. Beinat, S. Sobolevsky, P. Kazakopoulos, and C. Ratti. "Geo-Located Twitter as Proxy for Global Mobility Patterns." *Cartography and Geographic Information Science* 41, no. 3 (2014): 260–71.

Hawkins, E., and R. Sutton. "The Potential to Narrow Uncertainty in Regional Climate Predictions." *Bulletin of the American Meteorological Society* 90, no. 8 (2009): 1095–1107.

Haworth, B., and E. Bruce. "A Review of Volunteered Geographic Information for Disaster Management." *Geography Compass* 9, no. 5 (2015): 237–50.

Haworth, B., E. Bruce, and P. Middleton. "Emerging Technologies for Risk Reduction: Assessing the Potential Use of Social Media and VGI for Increasing Community Engagement." *Australian Journal of Emergency Management* 30, no. 3 (2015): 36–41.

Haythornthwaite, C., and L. Kendall. "Internet and Community." *American Behavioral Scientist* 53, no. 8 (2010): 1083–94. doi:10.1177/0002764209356242.

Healey, P. *Urban Complexity and Spatial Strategies: Towards a Relational Planning for our Times*. London: Routledge, 2007.

Hecht, B., and Stephens, M. "A Tale of Cities: Urban Biases in Volunteered Geographic Information." In *Proceedings of the Eighth International AAAI Conference on Weblogs and Social Media*. Palo Alto CA: AAAI Press, 2014.

Hecht, G. "Technology, Politics, and National Identity in France." In *Technologies of Power: Essays in Honor of Thomas Parke Hughes and Agatha Chipley Hughes*, edited by M. Thad Allen and G. Hecht, 145–74. Cambridge MA: MIT Press, 2001.

Helmreich, S. *Silicon Second Nature: Culturing Artificial Life in a Digital World*. Berkeley: University of California Press, 1998.

Hern, A. "Why Foursquare Should Be on Everyone's Phone." *Guardian*, May 26, 2014. http://www.theguardian.com/technology/2014/may/26/why-foursquare-isnt-just-the-timeline-spammer-you-thought-it-was-and-should-be-on-everyones-phone.

Hey, T., S. Tansley, K. Tolle, eds. *The Fourth Paradigm: Data-Intensive Scientific Discovery*. Redmond WA: Microsoft Research, 2009. http://www.openisbn.org/download/0982544200.pdf.

Hidalgo, C. "Saving Big Data from Big Mouths." *Scientific American*, April 29, 2014. http://www.scientificamerican.com/article/saving-big-data-from-big-mouths/.

Hilbert, M., and P. López. "The World's Technological Capacity to Store, Communicate, and Compute Information." *Science* 331 (February 10, 2011): 703–5.

Hiltz, S., J. Kushma, and L. Plotnick. "Use of Social Media by U.S. Public Sector Emergency Managers: Barriers and Wish Lists." In *Proceedings of the 11th International ISCRAM Conference*, edited by S. R. Hiltz, M. S. Pfaff, L. Plotnick, and P. C. Shih, 602–11. University Park: Pennsylvania State University, 2014. http://iscram2014.ist.psu.edu/sites/default/files/misc/proceedings/p11.pdf.

Historic Kansas City. "Our City: A Rich History and Culture." *Historic Kansas City*. http://www.historickansascity.org/?pages=kansas-city. Last accessed May 8, 2017.

Horton, R. "Offline: What Is Medicine's 5 Sigma?" *Lancet* 385 (April 11, 2015): 1380.

Howarth, D. *Discourse*. Buckingham UK: Open University Press, 2000.

Howe, J. "The Rise of Crowdsourcing." *Wired* 14, no. 6 (June 1, 2006). http://www.wired.com/2006/06/crowds/.

Howe, P., E. Markowitz, T. Lee, C. Ko, and A. Leiserowitz. "Global Perceptions of Local Temperature Change." *Nature Climate Change* 3, no. 4 (2013): 352–56. doi:10.1038/nclimate1768.

Hudson-Smith, A., M. Batty, A. Crooks, and R. Milton. "Mapping for the Masses: Accessing Web 2.0 through Crowdsourcing." *Social Science Computer Review* 27, no. 4 (2009): 524–38. doi:10.1177/0894439309332299.

Hudson-Smith, A., A. Crooks, M. Gibin, R. Milton, and M. Batty. "NeoGeography and Web 2.0: Concepts, Tools and Applications." *Journal of Location Based Services* 3, no. 2 (2009): 118–45. doi:10.1080/17489720902950366.

Hughes, A., and L. Palen. "The Evolving Role of the Public Information Officer: An Examination of Social Media in Emergency Management." *Journal of Homeland Security and Emergency Management* 9, no. 1 (2012).

Huizingh, E. "Open Innovation: State of the Art and Future Perspectives." *Technovation* 31, no. 1 (2011): 2–9.

Hulme, M. "Geographical Work at the Boundaries of Climate Change." *Transactions of the Institute of British Geographers* 33, no. 1 (2008): 5–11.

Humphreys, L., P. Gill, and B. Krishnamurthy. "How Much Is Too Much? Privacy Issues on Twitter." In *Proceedings of Conference of International Communication Association*. Singapore: International Communication Association, 2010.

Hyman, P. "'Small Data' Enabled Prediction of Obama's Win, Say Economists." *Communications of the ACM* 56, no. 5 (2013): 23–25.

Hyvärinen, O., and E. Saltikoff. "Social Media as a Source of Meteorological Observations." *Monthly Weather Review* 138, no. 8 (2010): 3175–84. doi:10.1175/2010MWR3270.1.

Infochimps. "CIOs and Big Data: What Your IT Team Wants You to Know." *Infochimps*, 2013. http://www.infochimps.com/resources/report-cios-big-data-what-your-it-team-wants-you-to-know-6/.

Information and Privacy Commissioner, Ontario. *The Seven Principles of Privacy by Design*, 2015. https://www.privacybydesign.ca/index.php/about-pbd/7-foundational-principles/.

Innes, M. "Control Creep." *Sociological Research Online* 6, no. 3 (2001). http://www.socresonline.org.uk/6/3/innes.html.

Jacobs, A. "The Pathologies of Big Data." *Communications of the ACM* 52, no. 8 (2009): 36–44.

Jaeger, P., S. Paquette, and S. Simmons. "Information Policy in National Political Campaigns: A Comparison of the 2008 Campaigns for President of the United States and Prime Minister of Canada." *Journal of Information Technology and Politics* 7, no. 1 (2010): 67–82.

James, R. "Out of the Box: Big Data Needs the Information Profession; The Importance of Validation." *Business Information Review* 31, no. 2 (2014): 118–21.

Johnson, R., and A. Onwuegbuzie. "Mixed Methods Research: A Research Paradigm Whose Time Has Come." *Educational Researcher* 33, no. 7 (2004): 14–26.

Johnston, R., R. Harris, K. Jones, D. Manley, C. Sabel, and W. Wang. "Mutual Misunderstanding and Avoidance, Misrepresentations and Disciplinary Politics:

Spatial Science and Quantitative Analysis in (United Kingdom) Geographical Curricula." *Dialogues in Human Geography* 4, no. 1 (2014): 3–25.

Jones, M. "Phase Space: Geography, Relational Thinking, and Beyond." *Progress in Human Geography* 33, no. 4 (2009): 487–506. doi:10.1177/0309132508101599.

Jungherr, A., P. Jürgens, and H. Schoen. "Why the Pirate Party Won the German Election of 2009 or the Trouble with Predictions." *Social Science Computer Review* 30, no. 2 (2012): 229–34.

Kahan, D. "Making Climate-Science Communication Evidence-Based—All the Way Down." In *Culture, Politics and Climate Change: How Information Shapes Our Common Future,* edited by D. Crow and M. Boykoff, 203–20. London: Routledge, 2013. http://papers.ssrn.com/sol3/papers.

Kaisler, S., F. Armour, J. Espinosa, and W. Money. 2013. "Big Data: Issues and Challenges Moving Forward." In *2013 46th Hawaii International Conference on System Sciences (HICSS),* 995–1004. doi:10.1109/HICSS.2013.645.

Kansas City Convention and Visitors Association. "Kansas City Is the City of Fountains." *Visit KC.* http://www.visitkc.com/this-is-kansas-city/overview/fountains/index.aspx.

Karimi, H., ed. *Big Data: Techniques and Technologies in Geoinformatics.* New York: Taylor and Francis, 2014.

Kauffman, S. *At Home in the Universe: The Search for Laws of Complexity.* London: Penguin, 1995.

Kaye, N., A. Hartley, and D. Hemming. "Mapping the Climate: Guidance on Appropriate Techniques to Map Climate Variables and Their Uncertainty." *Geoscientific Model Development Discussions* 4 (2011): 1875–1906.

Keen, S. *Debunking Economics: The Naked Emperor Dethroned.* London: Zed, 2011.

Kellens, W., T. Terpstra, and P. De Maeyer. "Perception and Communication of Flood Risks: A Systematic Review of Empirical Research." *Risk Analysis* 33, no. 1 (2012): 24–49. doi:10.1111/j.1539-6924.2012.01844.x.

Keller, C., M. Siegrist, and H. Gutscher. "The Role of the Affect and Availability Heuristics in Risk Communication." *Risk Analysis* 26, no. 3 (2006): 631–39. doi:10.1111/j.1539-6924.2006.00773.x.

Kelley, M. "The Emergent Urban Imaginaries of Geosocial Media." *GeoJournal* 78, no. 1 (2013): 181–203. doi:10.1007/s10708-011-9439-1.

———. "The Semantic Production of Space: Pervasive Computing and the Urban Landscape." *Environment and Planning A* 4, no. 4 (2014): 837–51.

———. "Urban Experience Takes an Informational Turn: Mobile Internet Usage and the Unevenness of Geosocial Activity." *GeoJournal* 79, no. 1 (2014): 15–29. doi:10.1007/s10708-013-9482-1.

Kelly, K. "We Are the Web." *Wired*, August 2005.

Kent, J., and H. Capello Jr. "Spatial Patterns and Demographic Indicators of Effective Social Media Content during the Horsethief Canyon Fire of 2012." *Cartography and Geographic Information Science* 40, no. 2 (2013): 78–89.

Kepes, Ben. "How to Kill Your Ecosystem: Twitter Pulls an Evil Move with Its Firehose." *Forbes.com*, April 11, 2015. http://www.forbes.com/sites/benkepes/2015/04/11/how-to-kill-your-ecosystem-twitter-pulls-an-evil-move-with-its-firehose/.

Kim Sang, E., and J. Bos. "Predicting the 2011 Dutch Senate Election Results with Twitter." *Proceedings of the 13th Conference of the European Chapter of the Association for Computational Linguistics*, 53–60. Stroudsburg PA: Association for Computational Linguistics, 2012.

King, L. *Statistical Analysis in Geography*. London: Prentice Hall, 1969.

Kingsbury, P., and J. P. Jones III. "Walter Benjamin's Dionysian Adventures on Google Earth." *Geoforum* 40 (2009): 502–13.

Kinsley, S. "Beyond the Screen: Methods for Investigating Geographies of Life 'Online.'" *Geography Compass* 7, no. 8 (2013): 540–55.

———. "Futures in the Making: Practices to Anticipate 'Ubiquitous Computing.'" *Environment and Planning A* 44, no. 7 (2012): 1554–69.

———. "The Matter of 'Virtual' Geographies." *Progress in Human Geography* 38, no. 3 (2014): 364–84.

Kirkpatrick, R. "Data Philanthropy: Public and Private Sector Data Sharing for Global Resilience." UN *Global Pulse* (blog), September 16, 2011. http://www.unglobalpulse.org/blog/data-philanthropy-public-private-sector-data-sharing-global-resilience.

Kitchin, R. "Big Data and Human Geography: Opportunities, Challenges and Risks." *Dialogues in Human Geography* 3, no. 3 (2013): 262–67.

———. "Big Data, New Epistemologies and Paradigm Shifts." *Big Data and Society* 1, no. 1 (2014). doi:10.1177/2053951714528481.

———. *The Data Revolution: Big Data, Open Data, Data Infrastructures and Their Consequences*. London: SAGE, 2014.

———. "Short Presentation on the Need for Critical Data Studies." *The Programmable City*, April 16, 2014. http://www.nuim.ie/progcity/2014/04/short-presentation-on-the-need-for-critical-data-studies/.

Kitchin, R., and M. Dodge. "Code and the Transduction of Space." *Annals of the Association of American Geographers* 95, no. 1 (2005): 162–80. doi:10.1111/j.1467-8306.2005.00454.x.

———. *Code/Space: Software and Everyday Life*. Cambridge MA: MIT Press, 2011.

Kitchin, R., and T. Lauriault. "Small Data in the Era of Big Data." *GeoJournal* 80, no. 4 (2015): 463–75.

Kleine, D. "Negotiating Partnerships, Understanding Power: Doing Action Research on Chilean Fairtrade Wine Value Chains." *Geographical Journal* 174, no. 2 (2008): 109–23.

Kluemper, D., H. Davison, X. Cao, and B. Wu. "Social Networking Websites and Personnel Selection." In *Employee Recruitment, Selection, and Assessment: Contemporary Issues for Theory and Practice*, edited by I. Nikolaou and J. Oostrom, 61–79. New York: Psychology Press, 2015.

Koops, B. "Forgetting Footprints, Shunning Shadows: A Critical Analysis of the 'Right to Be Forgotten' in Big Data Practice." SCRIPTed 8, no. 3 (2011): 229–56.

Korson, C. "Political Agency and Citizen Journalism: Twitter as a Tool of Evaluation." *Professional Geographer* 67, no. 3 (2015): 364–73.

Kounadi, O., and M. Leitner. "Why Does Geoprivacy Matter? The Scientific Publication of Confidential Data Presented on Maps." *Journal of Empirical Research on Human Research Ethics* 9, no. 4 (2014): 34–45.

Kruger, D. *Access Denied: Preventing Information Exclusion*. London: Demos, 1998.

Krumm, J. "A Survey of Computational Privacy." *Personal and Ubiquitous Computing* 13, no. 6 (2009): 391–99.

Kvasny, L. "Cultural (Re)production of Digital Inequality in a US Community Technology Initiative." *Information, Communication and Society* 9, no. 2 (2006): 160–81.

Kvasny, L., and M. Keil. "The Challenges of Redressing the Digital Divide: A Tale of Two US Cities." *Information Systems Journal* 16, no. 1 (2006): 23–53.

Kwan, M. "Gender and Individual Access to Urban Opportunities: A Study Using Space-Time Measures." *Professional Geographer* 51, no. 2 (1999): 210–27.

Kwan, M., and L. Knigge. "Guest Editorial." *Environment and Planning A* 38, no. 11 (2006): 1999–2002.

Lake, R. "Methods and Moral Inquiry." *Urban Geography* 35, no. 5 (2014): 657–68.

Laney, D. 3D *Data Management: Controlling Data Volume, Velocity and Variety*. Stamford CT: META Group, 2001. http://blogs.gartner.com/doug-laney/files/2012/01/ad949-3D-Data-Management-Controlling-Data-Volume-Velocity-and-Variety.pdf.

Lanier, J. *Who Owns the Future?* New York: Simon and Schuster, 2014.

———. *You Are Not a Gadget: A Manifesto*. London: Allen Lane, 2010.

Latour, B., and S. Woolgar. *Laboratory Life: The Construction of Scientific Facts*. Beverly Hills CA: SAGE, 1979.

Lauriault, T. "Data, Infrastructures and Geographical Imaginations." PhD thesis, Carleton University, 2012.

———. "Critical Analysis of the Irish Big Data Skills Report." *The Programmable City*, May 28, 2014. http://www.nuim.ie/progcity/2014/05/big-data-series-critical -analysis-of-the-irish-big-data-skills-report/.

Lauriault, T., B. Craig, D. Taylor, and P. Pulsifier. "Today's Data Are Part of Tomorrow's Research: Archival Issues in the Sciences." *Archivaria* 64 (2007): 123–79.

LaValle, S., E. Lesser, R. Shockey, M. Hopkins, and N. Kruschwitz. "Big Data, Analytics, and the Path from Insights to Value." MIT *Sloan Management Review* 52, no. 2 (2011): 21–31.

Lawson, T. *Economics and Reality*. London: Routledge, 1997.

Lazer, D., R. Kennedy, G. King, and A. Vespignani. "The Parable of Google Flu: Traps in Big Data Analysis." *Science* 343, no. 6176 (2014): 1203–5.

Lee, N. *Facebook Nation: Total Information Awareness*. 2nd ed. New York: Springer, 2014.

Leetaru, K., S. Wang, G. Cao, A. Padmanabhan, and E. Shook. "Mapping the Global Twitter Heartbeat: The Geography of Twitter." *First Monday* 18, no. 5 (2013). http://dx.doi.org/10.5210/fm.v18i5.4366.

Lefebvre, H. Henri. *The Production of Space*. Oxford: Blackwell, 1991.

Lehdonvirta, V., and M. Ernkvist. *Knowledge Map of the Virtual Economy*. Washington DC: World Bank, 2011. http://www.infodev.org/en/Document.1056.pdf.

Leibowitz, Barry. "Philadelphia Blogger Charged for Threatening Facebook Pages." *Crimesider*, CBSNews.com, September 12, 2012. http://www.cbsnews.com/news /philadelphia-blogger-charged-for-threatening-facebook-pages/.

Leiserowitz, A. "Communicating the Risks of Global Warming: American Risk Perceptions, Affective Images, and Interpretive Communities." In *Creating a Climate for Change: Communicating Climate Change and Facilitating Social Change*, edited by S. Moser and L. Dilling, 44–63. Cambridge: Cambridge University Press, 2007.

Leonelli, S. "What Difference Does Quantity Make? On the Epistemology of Big Data in Biology." *Big Data and Society* 1, no. 1 (2014). doi:2053951714534395.

Lerman, J. "Big Data and Its Exclusions." *Stanford Law Review Online* 66 (September 2013): 55.

Leszczynski, A. "On the Neo in Neogeography." *Annals of the Association of American Geographers* 104, no. 1 (2014): 60–79. doi:10.1080/00045608.2013.846159.

———. "Quantitative Limits to Qualitative Engagements: GIS, Its Critics, and the Philosophical Divide." *Professional Geographer* 61, no. 3 (2009): 350–65.

———. "Situating the Geoweb in Political Economy." *Progress in Human Geography* 36, no. 1 (2012): 72–89.

———. "Spatial Big Data and Anxieties of Control." *Environment and Planning D* 33, no. 6 (2015): 965–84.

Leszczynski, A., and S. Elwood. "Feminist Geographies of New Spatial Media."
Canadian Geographer/Le Géographe Canadien 59, no. 1 (2015): 12–28. doi:10.1111
/cag.12093.

Leszczynski, A., and M. Wilson. "Guest Editorial: Theorizing the Geoweb." *GeoJournal* 78, no. 6 (2013): 915–19. http://link.springer.com/article/10.1007/s10708
-013-9489-7.

Letouzé, E. "Big Data for Development: Challenges and Opportunities." UN *Global
Pulse* (blog), May 2012. http://www.unglobalpulse.org/projects/BigDatafor
Development.

———. "Can Big Data from Cellphones Help Prevent Conflict?" *Global Observatory*
(blog), November 8, 2012. http://theglobalobservatory.org/analysis/380-can-big
-data-from-cellphones-help-prevent-conflict.html.

Levy, S. *Hackers: Heroes of the Computer Revolution.* Garden City NY: Anchor
Press, 1984.

Lewandowsky, S, J. Risbey, M. Smithson, B. Newell, and J. Hunter. "Scientific Uncertainty and Climate Change: Part I. Uncertainty and Unabated Emissions." *Climatic
Change* 124, no. 1–2 (2014): 21–37. doi:10.1007/s10584-014-1082-7.

Li, L., and M. Goodchild. "The Role of Social Networks in Emergency Management:
A Research Agenda." *International Journal of Information Systems for Crisis Response
and Management* 2, no. 4 (2010): 49–59.

Li, L., M. Goodchild, and B. Xu. "Spatial, Temporal, and Socioeconomic Patterns
in the Use of Twitter and Flickr." *Cartography and Geographic Information Science*
40, no. 2 (2013): 61–77.

Light, J. *From Warfare to Welfare: Defense Intellectuals and Urban Problems in Cold
War America.* Baltimore: Johns Hopkins University Press, 2003.

Lin, J., S. Amini, J. I. Hong, N. Sadeh, J. Lindqvist, and J. Zhang. "Expectation and
Purpose: Understanding Users' Mental Models of Mobile App Privacy through
Crowdsourcing." In *Proceedings of the 2012 ACM Conference on Ubiquitous Computing,* 501–10. New York: ACM, 2012. doi:10.1145/2370216.2370290.

Lin, W. "Situating Performative Neogeography: Tracing, Mapping, and Performing
'Everyone's East Lake.'" *Environment and Planning A* 45, no. 1 (2013): 37–54.

———. "Volunteered Geographic Information and Networked Publics? Politics of
Everyday Mapping and Spatial Narratives." *GeoJournal* 78, no. 6 (2013): 949–65.
doi:10.1007/s10708-013-9490-1.

Lindqvist, J., J. Cranshaw, J. Wiese, J. Hong, and J. Zimmerman. "I'm the Mayor of
My House: Examining Why People Use Foursquare—a Social-Driven Location
Sharing Application." *Proceedings of the 2011 Conference on Human Factors in Computing Systems,* 2409–18. New York: ACM, 2011. doi:10.1145/1978942.1979295.

Lindsay, C. "Employability, Services for Unemployed Job Seekers and the Digital Divide." *Urban Studies* 42, no. 2 (2005): 325–39. doi:10.1080 /0042098042000316173.

Lini, S., "Employers Eyeing Twitter, Social Media." *Privacy Journal* 38, no. 5 (2012): 1–7.

Liu, S., J. Fraustino, and Y. Jin. "How Disaster Information Form, Source, Type, and Prior Disaster Exposure Affect Public Outcomes: Jumping on the Social Media Bandwagon?" *Journal of Applied Communication Research* 43, no. 1 (2015): 44–65.

Liu, S., and L. Palen. "The New Cartographers: Crisis Map Mashups and the Emergence of Neogeographic Practice." *Cartography and Geographic Information Science* 37, no. 1 (2010): 69–90.

Liu, S., L. Palen, J. Sutton, A. Hughes, and S. Vieweg. "Citizen Photojournalism during Crisis Events." *Citizen Journalism: Global Perspectives*, edited by S. Allan and E. Thorsen, 43–63. New York: Peter Lang, 2009.

Loeffler, C. "Privacy Issues in Social Media." *IP Litigator* 18, no. 5 (2012): 12–18.

Loewenstein, G., E. Weber, C. Hsee, and N. Welch. "Risk as Feelings." *Psychological Bulletin* 127, no. 2 (2001): 267–86. doi:10.1037/0033-2909.127.2.267.

Lohrenz, M., J. Trafton, M. Beck, and M. Gendron. "A Model of Clutter for Complex, Multivariate Geospatial Displays." *Human Factors: The Journal of the Human Factors and Ergonomics Society* 51, no. 1 (2009): 90–101. doi:10.1177/0018720809333518.

Lorenzoni, I., S. Nicholson-Cole, and L. Whitmarsh. "Barriers Perceived to Engaging with Climate Change among the UK Public and Their Policy Implications." *Global Environmental Change* 17, no. 3–4 (2007): 445–59.

Lu, J., and D. Li. "Bias Correction in a Small Sample from Big Data." *IEEE Transactions on Knowledge and Data Engineering* 25, no. 11 (2013): 2658–63.

Lu, Y. "Trouble in Start-Up Land: Silicon Valley's Youth Problem." *New York Times Magazine*, March 12, 2014.

Lyon, D. *Surveillance Studies: An Overview.* Cambridge MA: Polity, 2007.

MacEachren, A., A. Robinson, S. Hopper, S. Gardner, R. Murray, M. Gahegan, and E. Hetzler. "Visualizing Geospatial Information Uncertainty: What We Know and What We Need to Know." *Cartography and Geographic Information Science* 32, no. 3 (2005): 139–60.

MacEachren, A., A. Robinson, A. Jaiswal, S. Pezanowski, A. Savelyev, J. Blanford, and P. Mitra. "Geo-Twitter Analytics: Applications in Crisis Management." Paper presented at the 25th International Cartographic Conference, Paris, France, 2011. https://www.geovista.psu.edu/publications/2011/MacEachren_ICC_2011.pdf.

Maguire, D., and P. Longley. "The Emergence of Geoportals and Their Role in Spatial Data Infrastructures." *Computers, Environment and Urban Systems* 29, no. 1 (2005): 3–14.

Maibach, E., C. Roser-Renouf, and A. Leiserowitz. *Global Warming's Six Americas 2009: An Audience Segmentation Analysis.* New Haven CT: Yale Project on Climate Change, School of Forestry and Environmental Studies, Yale University; Fairfax VA: Center for Climate Change Communication, Department of Communication, George Mason University, 2009.

Malecki, E. J., and B. Moriset. *The Digital Economy: Business Organization, Production Processes, and Regional Developments.* New York: Routledge, 2008.

Manovich, L. "Trending: The Promise and the Challenges of Big Social Data." *Manovich.net,* 2011. http://manovich.net/index.php/projects/trending-the -promises-and-the-challenges-of-big-social-data.

Manson, S. "Simplifying Complexity: A Review of Complexity Theory." *Geoforum* 32, no. 3 (2001): 405–14.

Manson, S., and D. O'Sullivan. "Complexity Theory in the Study of Space and Place." *Environment and Planning A* 38, no. 4 (2006): 677–92.

Manyika, J., M. Chiu, B. Brown, J. Bughin, R. Dobbs, C. Roxburgh, and A. Hung Byers. *Big Data: The Next Frontier for Innovation, Competition, and Productivity.* New York: McKinsey Global Institute, 2011.

Markowsky, G., "In Praise of Small Data." *2014 Academy of Science and Engineering (ASE) Bigdata/SocialCom/Cyber Security Conference Workshop Papers,* Stanford University, May 27–31, 2014. http://ase360.org/handle/123456789/136.

Martin, D. "Constructing Place: Cultural Hegemonies and Media Images of an Inner-City Neighborhood." *Urban Geography* 21, no. 5 (2000): 380–405.

Marz, N., and J. Warren. *Big Data: Principles and Best Practices of Scalable Realtime Data Systems.* MEAP ed. Shelter Island NY: Manning, 2012.

Mashey, J. "Big Data and the Next Wave of InfraStress." Presentation at Computer Science Division Seminar, University of California, Berkeley, April 25, 1998.

Massey, D. *Space, Place, and Gender.* Minneapolis: University of Minnesota Press, 1994.

Massey, D., A. Gross, and K. Shibuya. "Migration, Segregation, and the Geographic Concentration of Poverty." *American Sociological Review* 59 (1994): 425–45.

Mattingly, D., and K. Falconer-Al-Hindi. "Should Women Count? A Context for the Debate." *Professional Geographer* 47, no. 4 (1995): 427–35.

Mayer-Schönberger, V., and K. Cukier. *Big Data: A Revolution That Will Transform How We Live, Work, and Think.* New York: Houghton Mifflin Harcourt, 2013.

McChesney, R. *Digital Disconnect: How Capitalism Is Turning the Internet against Democracy.* New York: New Press, 2013.

McLafferty, S. "Counting for Women." *Professional Geographer* 47, no. 4 (1995): 436–42.

Meeker, M. "Top Mobile Internet Trends." KPCB (blog), February 10, 2011. http://www.slideshare.net/kleinerperkins/kpcb-top-10-mobile-trends-feb-2011.

Meeker, M., and L. Wu. "Internet Trends." KPCB (blog), May 29, 2013. http://www.kpcb.com/blog/2013-internet-trends.

Meier, P. "Debating the Value of Tweets for Disaster Response (Intelligently)." *iRevolution* (blog), December 17, 2012. http://irevolution.net/2012/12/17/debating-tweets-disaster/.

———. *Digital Humanitarians: How Big Data Is Changing the Face of Humanitarian Response.* Boca Raton FL: CRC Press, 2015.

———. "Human Computation for Disaster Response. In *Handbook of Human Computation,* edited by P. Michelucci, 95–104. New York: Springer, 2013. http://dx.doi.org/10.1007/978-1-4614-8806-4_11.

———. "Humanitarians in the Sky: Using UAVs for Disaster Response." *iRevolution* (blog), June 25, 2014. http://irevolution.net/2014/06/25/humanitarians-in-the-sky-using-uavs-for-disaster-response/.

———. "What Is Big (Crisis) Data?" *iRevolution* (blog), June 27, 2013. http://irevolution.net/2013/06/27/what-is-big-crisis-data/.

———. "What Was Novel about Social Media Use during Hurricane Sandy?" *iRevolution* (blog), October 31, 2012. http://irevolution.net/2012/10/31/hurricane-sandy/.

Melillo, J., T. Richmond, and G. Yohe, eds. *Climate Change Impacts in the United States: The Third National Climate Assessment.* U.S. Global Change Research Program, 2014. http://www.globalchange.gov/browse/reports/climate-change-impacts-united-states-third-national-climate-assessment-0.

Mervyn, K., and D. Allen. "Sociospatial Context and Information Behavior: Social Exclusion and the Influence of Mobile Information Technology." *Journal of the American Society for Information Science and Technology* 63, no. 6 (2012): 1125–41. doi:10.1002/asi.22626.

Mesch, G., and I. Talmud. "Internet Connectivity, Community Participation, and Place Attachment: A Longitudinal Study." *American Behavioral Scientist* 53, no. 8 (2010): 1095–1110. doi:10.1177/0002764209356243.

Meyen, M., S. Pfaff-Rudiger, K. Dudenhoffer, and J. Huss. "The Internet in Everyday Life: A Typology of Internet Users." *Media Culture and Society* 32, no. 5 (2010): 873–82. doi:10.1177/0163443710374792.

Middleton, J. "Email Urges 'New Zealander' for Census." *New Zealand Herald,* March 1, 2006. http://www.nzherald.co.nz/nz/news/article.cfm?c_id=1andobjectid=10370546.

Miles, M., and A. Huberman, eds. *Qualitative Data Analysis: An Expanded Sourcebook.* 2nd ed. Thousand Oaks CA: SAGE, 1994.

Miller, D. "Could the Internet Defetishise the Commodity?" *Environment and Planning D: Society and Space* 21, no. 3 (2003): 359–72.

Miller, G. "Social Scientists Wade into the Tweet Stream." *Science* 333 (2011): 1814–15.

Miller, H., and M. Goodchild. "Data-Driven Geography." *GeoJournal* 80, no. 4 (2014): 449–61. http://link.springer.com/10.1007/s10708-014-9602-6.

Miller, H., and J. Han J, eds. *Geographic Data Mining and Knowledge Discovery.* 2nd ed. Boca Raton FL: CRC Press, 2009.

Miller, P. "Digital Divide: Segregation Is Alive and Well in Social Media." *Advertising Age* 82, no. 6 (2011).

Mills, C. *The Sociological Imagination.* Oxford: Oxford University Press, 2000.

Minelli, M., M. Chambers, and A. Dhiraj. *Big Data, Big Analytics.* Hoboken NJ: Wiley, 2013.

Mitchell, M. *Complexity: A Guided Tour.* New York: Oxford University Press, 2008.

Mitra, A. "Creating a Presence on Social Networks via Narbs." *Global Media Journal* 9, no. 16 (2010).

Mizen, H., C. Dolbear, and G. Hart. "Ontology Ontogeny: Understanding How an Ontology Is Created and Developed." In *GeoSpatial Semantics: First International Conference*, Lecture Notes in Computer Science 3799, edited by M. Rodriguez, I. Cruz, M. Egenhofer, and S. Levashkin, 15–29. Berlin: Springer, 2005.

Monmonier, M. *Spying with Maps: Surveillance Technologies and the Future of Privacy.* Chicago: University of Chicago Press, 2008.

Monroe, B. "The Five Vs of Big Data Political Science: Introduction to the Virtual Issue on Big Data in Political Science." *Political Analysis Virtual Issue* 4 (2013).

Morozov, E. *To Save Everything, Click Here: The Folly of Technological Solutionism.* New York: Public Affairs, 2013.

Morse, J. "Principles of Mixed Methods and Multimedia Research Design." In *Handbook of Mixed Methodology in Social and Behavioral Research*, edited by A. Tashakkori and C. Teddlie, 189–208. Thousand Oaks CA: Sage, 2003.

Moser, S. "Communicating Climate Change: History, Challenges, Process and Future Directions." *Wiley Interdisciplinary Reviews: Climate Change* 1, no. 1 (2010): 31–53.

Mossberger, K., C. J. Tolbert, D. Bowen, and B. Jimenez. "Unraveling Different Barriers to Internet Use: Urban Residents and Neighborhood Effects." *Urban Affairs Review* 48, no. 6 (2012): 771–810. doi:10.1177/1078087412453713.

Murthy, D. "Digital Ethnography: An Examination of the Use of New Technologies for Social Research." *Sociology* 42, no. 5 (2008): 837–55.

Narula, S. "The Real Problem with a Service Called 'Ghetto Tracker.'" *The Atlantic*, September 6, 2013.

Nast, H. J. "Women in the Field: Critical Feminist Methodologies and Theoretical Perspectives." *Professional Geographer* 46, no. 1 (1994): 54–66. doi:10.1111/j.0033 -0124.1994.00054.x.

National Science Foundation (NSF). "Critical Techniques, Technologies and Methodologies for Advancing Foundations and Applications of Big Data Sciences and Engineering (BIGDATA)." Arlington VA, 2012. http://www.nsf.gov/funding /pgm_summ.jsp?pims_id=504767.

Naughton, J. "Technology Is a Double-Edged Sword." *Guardian*, December 22, 2012. http://www.theguardian.com/technology/2012/dec/23/creative-destruction -john-naughton-networker.

Nelson, S. "Big Data: The Harvard Computers." *Nature* 455 (2008): 36–37.

Newman, G., A. Wiggins, A. Crall, E. Graham, S. Newman, and K. Crowston. "The Future of Citizen Science: Emerging Technologies and Shifting Paradigms." *Frontiers in Ecology and the Environment* 10, no. 6 (2012): 298–304. doi:10.1890/110294.

Nicholson-Cole, S. "Representing Climate Change Futures: A Critique on the Use of Images for Visual Communication." *Computers, Environment and Urban Systems* 29, no. 3 (2005): 255–73. doi:10.1016/j.compenvurbsys.2004.05.002.

Nickerson, D., and T. Rogers. "Political Campaigns and Big Data." *Journal of Economic Perspectives* 28, no. 2 (2014): 51–74.

Nightingale, A. "A Feminist in the Forest: Situated Knowledges and Mixing Methods in Natural Resource Management." *ACME: An International E-Journal for Critical Geographies* 2, no. 1 (2003): 77–90.

Nisbet, M. "Communicating Climate Change: Why Frames Matter for Public Engagement." *Environment: Science and Policy for Sustainable Development* 51, no. 2 (2009): 12–23.

Nissenbaum, H. *Privacy in Context: Technology, Policy and the Integrity of Social Life.* Stanford CA: Stanford Law Books, 2010.

Nittel, S., L. Bodum, K. Clarke, M. Gould, P. Raposo, J. Sharma, and M. Vasardani. "Emerging Technological Trends Likely to Affect GIScience in the Next 20 Years." *Advancing Geographic Information Science*, September 20, 2015. https:// www.researchgate.net/profile/Silvia_Nittel/publication/281973116_Emerging _Technological_Trends_likely_to_Affect_GIScience_in_the_Next_20_Years /links/56003fa908aeafc8ac8c50c3.pdf.

Nitze, A. "Evaluation of JavaScript Quality Issues and Solutions for Enterprise Application Development." In *Software Quality: Software and Systems Quality in Distrib-*

uted and Mobile Environments, edited by D. Winkler, S. Biffl, and J. Bergsmann, 108–19. New York: Springer International, 2015.

Niven, D., J. Sauer, G. Butcher, and W. Link. "Christmas Bird Count Provides Insights into Population Change in Land Birds That Breed in the Boreal Forest." *American Birds* 58 (2004): 10–20.

Nurse, J. "Exploring Risks to Identity Security and Privacy in Cyberspace." ACM *Magazine: Crossroads* 21, no. 3 (2015). doi:10.1145/2730912.

Ockwell, D., L. Whitmarsh, and S. O'Neill. "Reorienting Climate Change Communication for Effective Mitigation: Forcing People to Be Green or Fostering Grass-Roots Engagement?" *Science Communication* 30, no. 3 (2009): 305–27. doi:10 .1177/1075547008328969.

O'Connor, B., R. Balasubramanyan, B. Routledge, and N. Smith. "From Tweets to Polls: Linking Text Sentiment to Public Opinion Time Series." *Proceedings of the International AAAI Conference on Weblogs and Social Media*, 122–29. Washington DC: AAAI, 2010.

Office of the Auditor General of Canada. "Chapter 7: Documentary Heritage of the Government of Canada—Library and Archives Canada." *Report of the Auditor General*. Ottawa: Minister of Public Works and Government Services, 2014. http:// www.oag-bvg.gc.ca/internet/docs/parl_oag_201411_07_e.pdf.

Okolloh, O. "Ushahidi, or 'Testimony': Web 2.0 Tools for Crowdsourcing Crisis Information." *Participatory Learning and Action* 59, no. 1 (2009): 65–70.

Olafsson, G. *Humanitarian Response in the Age of Mass Collaboration and Networked Intelligence*. Washington DC: Woodrow Wilson International Center for Scholars, 2012. http://wilsoncenter.org/sites/default/files/VisionfortheFuture _PositionPaper_0.pdf.

Olfat, H., S. Soltanieh, A. Rajabifard, H. Senot, and I. Williamson. "Spatial Metadata Automation: A Key to Spatially Enabling Platform." *International Journal of Spatial Data Infrastructures Research* 7 (2012): 173–95.

Oliveira, R., I. Cardoso, J. Barbosa, C. da Costa, and M. Prado. "An Intelligent Model for Logistics Management Based on Geofencing Algorithms and RFID Technology." *Expert Systems with Applications* 42, no. 15 (2015): 6082–97.

Onsrud, H., and J. Pinto. "Diffusion of Geographic Information Innovations." *International Journal of Geographical Information Science* 5, no. 4 (1991): 447–67.

Open Data Center Alliance. *Big Data Consumer Guide*. Portland OR, 2012. https:// www.opendatacenteralliance.org/docs/Big_Data_Consumer_Guide_Rev1.0.pdf.

Openshaw, S. "Two Explanatory Space-Time-Attribute Pattern Analysers Relevant to GIS." In *Spatial Analysis and GIS*, edited by A. Fotheringham and P. Rogerson, 83–104. London: Taylor and Francis, 1994.

Oppenheimer, M., B. O'Neill, M. Webster, and S. Agrawala. "The Limits of Consensus." *Science* 317, no. 5844 (2007): 1505–6. doi:10.1126/science.1144831.

O'Reilly, T. "The Creep Factor: How to Think about Big Data and Privacy." *Forbes*, March 6, 2014. http://www.forbes.com/sites/oreillymedia/2014/03/06/the-creep-factor-how-to-think-about-big-data-and-privacy/.

———. "What Is Web 2.0: Design Patterns and Business Models for the Next Generation of Software." *O'Reilly Media*, September 30, 2005. http://www.oreilly.com/pub/a/web2/archive/what-is-web-20.html.

Oreskes, N., and E. Conway. *Merchants of Doubt: How a Handful of Scientists Obscured the Truth on Issues from Tobacco Smoke to Global Warming.* New York: Bloomsbury Publishing USA, 2010.

O'Sullivan, D. "Complexity Science and Human Geography." *Transactions of the Institute of British Geographers* 29, no. 3 (2004): 282–95.

———. "Geographical Information Science: Time Changes Everything." *Progress in Human Geography* 29, no. 6 (2005): 749–56.

O'Sullivan, D., and Perry, G. *Spatial Simulation: Exploring Pattern and Process.* Chichester UK: Wiley-Blackwell, 2013.

Palen, L., S. Vieweg, and K. Anderson. "Supporting 'Everyday Analysts' in Safety- and Time-Critical Situations." *Information Society* 27 (2011): 52–62.

Pariser, E. *The Filter Bubble: What the Internet Is Hiding from You.* New York: Penguin Press, 2011.

Park, R. E., E. W. Burgess, and R. D. McKenzie, eds. *The City.* Chicago: University of Chicago Press, 1967.

Parker, I. *Discourse Dynamics: Critical Analysis for Social and Individual Psychology.* London: Routledge, 1992.

Parker, S., E. Uprichard, and R. Burrows. "Class Places and Place Classes: Geodemographics and the Spatialization of Class." *Information, Communication and Society* 10, no. 6 (2007): 901–20. doi:10.1080/13691180701751122.

Parkinson, H. J. "Runner Uses Nike+ App to Draw Penises." *Guardian*, August 6, 2014. http://www.theguardian.com/technology/2014/aug/06/runner-nike-san-francisco-penis.

Pasquale, F. *The Black Box Society: The Secret Algorithms That Control Money and Information.* Cambridge MA: Harvard University Press, 2015.

Paul, A., and A. Bruns. "Usability of Small Crisis Data Sets in the Absence of Big Data." In *Proceedings of the 2013 International Conference on Information, Business and Education Technology*, edited by Ariwa, Ezendu, Zhao, Wenbing, and Gandhi, Meenakshi, 718–21. Beijing: Atlantis Press, 2013. http://eprints.qut.edu.au/58830/.

Paul, M., and M. Drezde. "You Are What You Tweet: Analyzing Twitter for Public Health." *Proceedings of the Fifth International AAAI Conference on Weblogs and Social Media*. Palo Alto CA: AAAI, 2011. http://www.aaai.org/ocs/index.php/ICWSM /ICWSM11/paper/view/2880/3264.

Pavlovskaya, M., and K. St. Martin. "Feminism and Geographic Information Systems: From a Missing Object to a Mapping Subject." *Geography Compass* 1, no. 3 (2007): 583–606.

Pearce, K., and R. Rice. "Digital Divides from Access to Activities: Comparing Mobile and Personal Computer Internet Users." *Journal of Communication* 63, no. 4 (2013): 721–44. doi:10.1111/jcom.12045.

Perez, S. "Following Twitter Suspension, WeKnowYourHouse Returns, Continues to Post Twitter Users' Addresses, Home Photos." *TechCrunch*, August 16, 2012. http:// techcrunch.com/2012/08/16/following-twitter-suspension-weknowyourhouse -returns-continues-to-post-twitter-users-addresses-home-photos/.

Pérez-Montoro, M., and J. Díaz Nafría. "Data." In *Glossary of Concepts, Metaphors, Theories and Problems Concerning Information,* edited by J. Díaz Nafría, M. Pérez-Montoro, and F. Alemany. León, Spain: Universidad de León, 2010. http:// glossarium.bitrum.unileon.es/Home/dato/data.

Peters, B. "The Big Data Gold Rush." *Forbes*, June 21, 2012. http://www.forbes.com /sites/bradpeters/2012/06/21/the-big-data-gold-rush/.

Pew Research Center. "Mobile Fact Sheet." *Internet and Technology*, January 12, 2017. http://www.pewinternet.org/fact-sheet/mobile/.

Pickles, J. *Ground Truth: The Social Implications of Geographic Information Systems.* New York: Guilford Press, 1995.

Pidgeon, N., and B. Fischhoff. "The Role of Social and Decision Sciences in Communicating Uncertain Climate Risks." *Nature Climate Change* 1, no. 1 (2011): 35–41. doi:10.1038/nclimate1080.

Pierce, J., D. Martin, and J. Murphy. "Relational Place-Making: the Networked Politics of Place." *Transactions of the Institute of British Geographers* 36, no. 1 (2011): 54–70. doi:10.1111/j.1475-5661.2010.00411.x.

Pigliucci, M. "The End of Theory in Science?" *EMBO Reports* 10, no. 6 (2009): 534.

Pink, S. "Visual Ethnography and the Internet: Visuality, Virtuality and the Spatial Turn." In *Advances in Visual Methodology*, edited by S. Pink, 113–30. London: SAGE, 2012.

Pither, K. *Dark Days: The Story of Four Canadians Tortured in the Name of Fighting Terror.* Toronto: Viking Canada, 2008.

Plummer, P., and E. Sheppard. "Must Emancipatory Economic Geography Be Qualitative?" *Antipode* 33, no. 2 (2001): 194–99.

Podesta, J., and J. Holdren. "Climate Data Initiative Launches with Strong Public and Private Sector Commitments—The White House." The White House, March 19, 2014. https://obamawhitehouse.archives.gov/blog/2014/03/19/climate-data -initiative-launches-strong-public-and-private-sector-commitments.

Pollock, R. "Forget Big Data, Small Data Is the Real Revolution." *Guardian*, April 25, 2013. http://www.theguardian.com/news/datablog/2013/apr/25/forget-big -data-small-data-revolution.

Poon, J. "Quantitative Methods: Past and Present." *Progress in Human Geography* 28, no. 6 (2004): 807–14.

Poovey, M. *A History of the Modern Fact: Problems of Knowledge in the Sciences of Wealth and Society.* Chicago: University of Chicago Press, 1998.

Popper, B., and E. Hamburger. "Meet Swarm: Foursquare's Ambitious Plan to Split Its App in Two." *The Verge*, May 1, 2014. http://www.theverge.com/2014/5/1 /5666062/foursquare-swarm-new-app.

Porter, T. *The Rise of Statistical Thinking.* Princeton NJ: Princeton University Press, 1986.

Postill, J., and S. Pink. "Social Media Ethnography: The Digital Researcher in a Messy Web." *Media International Australia, Incorporating Culture and Policy* 145 (2012): 124–32.

Postman, N. *Technopoly.* New York: Vintage, 1993.

Pourabbas, E., ed. *Geographical Information Systems: Trends and Technologies.* Boca Raton FL: CRC Press, 2014.

Press, G. "6 Predictions for the $125 Billion Big Data Analytics Market in 2015." *Forbes*, December 11, 2014. http://www.forbes.com/sites/gilpress/2014/12/11 /6-predictions-for-the-125-billion-big-data-analytics-market-in-2015/.

Primack, R., and A. Miller-Rushing. "Uncovering, Collecting, and Analyzing Records to Investigate the Ecological Impacts of Climate Change: A Template from Thoreau's Concord." *BioScience* 62, no. 2 (2012): 170–81. doi:10.1525/bio.2012.62.2.10.

Punathambekar, A., and A. Kavada. "Debating Big Data." *Media, Culture and Society* 37, no. 7 (2015): 1076–77.

Pylas, P. "Why Europe's Facebook Users Have Cause for Privacy Concerns." Associated Press, October 6, 2015. http://www.ctvnews.ca/sci-tech/why-europe-s -facebook-users-have-cause-for-privacy-concerns-1.2596863.

Raley, R. "Dataveillance and Countervailance." In *"Raw Data" Is an Oxymoron*, edited by L. Gitelman, 121–46. Cambridge MA: MIT Press, 2013.

Rast, E. "Context as Assumptions." In *Epistemology, Context, and Formalism*, edited by F. Lihoreau and M. Rebuschi, 9–31. Cham, Switzerland: Springer International, 2014.

Raymond, E. *The Cathedral and the Bazaar: Musings on Linux and Open Source by an Accidental Revolutionary.* Cambridge MA: O'Reilly Media, 2001.

Reidenberg, J., R. Gellman, J. Debelak, A. Elewa, and N. Liu. *Privacy and Missing Persons in Natural Disasters: Striking the Right Balance between Rights and Needs.* Washington DC: Woodrow Wilson International Center for Scholars, 2013. https://www.wilsoncenter.org/sites/default/files/Privacy_MissingPersons_FINAL.pdf.

Research Councils UK. "Big Data." N.d. http://www.rcuk.ac.uk/research/infrastructure/big-data/.

Retchless, D. "Communicating Climate Change: Spatial Analog versus Color-Banded Isoline Maps with and without Accompanying Text." *Cartography and Geographic Information Science* 40, no. 1 (2014): 55–74.

Rial, N. "The Power of Big Data in Europe." *New Europe,* May 24, 2013. http://www.neurope.eu/article/power-big-data-europe.

Ribes, D. and S. Jackson. "Data Bite Man: The Work of Sustaining Long-Term Study." In *"Raw Data" Is an Oxymoron,* edited by L. Gitelman, 147–66. Cambridge MA: MIT Press, 2013.

Richardson, K., P. Cilliers, and M. Lissack. "Complexity Science: A 'Gray' Science for the 'Stuff in Between.'" *Emergence* 3, no. 2 (2001): 6–18.

Richardson, M., and P. Domingos. "Mining Knowledge-Sharing Sites for Viral Marketing." In *Proceedings of the Eighth ACM SIGKDD International Conference on Knowledge Discovery and Data Mining,* edited by O. Zaïane, R. Goebel, D. Hand, D. Keim, and R. Ng, 61–70. New York: Association of Computing Machinery, 2002.

Ricker, B., N. Hedley, and S. Daniel. "Fuzzy Boundaries: Hybridizing Location-Based Services, Volunteered Geographic Information and Geovisualization Literature." *Geography Compass* 8, no. 7 (2014): 490–504.

Risen, T. "How the Safe Harbor Law Threatens U.S. Businesses." *U.S. News and World Report,* October 5, 2015. http://www.usnews.com/news/articles/2015/10/05/rejecting-safe-harbor-law-threatens-us-and-eu-businesses.

Robson, E. *Responding to Liability: Evaluating and Reducing Tort Liability for Digital Volunteers.* Washington DC: Woodrow Wilson International Center for Scholars, 2012. http://www.wilsoncenter.org/sites/default/files/responding_to_liability_1.pdf.

Roche, S., E. Propeck-Zimmermann, and B. Mericskay. "GeoWeb and Crisis Management: Issues and Perspectives of Volunteered Geographic Information." *GeoJournal* 78, no. 1 (2011): 1–20.

Rocheleau, D. "Maps, Numbers, Text, and Context: Mixing Methods in Feminist Political Ecology." *Professional Geographer* 47, no. 4 (1995): 458–66. doi:10.1111/j.0033-0124.1995.458_h.x.

Romero, Ric. "Are Insurance Companies Spying on Your Facebook Page?" *Eyewitness News/ABC7*, November 7, 2011. http://abc7.com/archive/8422388/.

Rose, G. *Feminism and Geography: The Limits of Geographical Knowledge*. Minneapolis: University of Minnesota Press, 1993.

――――. "Situating Knowledges: Positionality, Reflexivities and Other Tactics." *Progress in Human Geography* 21, no. 3 (1997): 305–20. doi:10.1191/030913297673302122.

Rosenberg, D. "Data before the Fact." In *"Raw Data" Is an Oxymoron*, edited by L. Gitelman, 15–40. Cambridge, MA: MIT Press, 2013

Rösler, R., and T. Liebig. "Using Data from Location Based Social Networks for Urban Activity Clustering." In *Geographic Information Science at the Heart of Europe, Lecture Notes in Geoinformation and Cartography*, edited by D. Vandenbroucke, B. Bucher, and J. Crompvoets, 55–72. Cham, Switzerland: Springer International, 2013.

Roth, R. "The Impact of User Expertise on Geographic Risk Assessment under Uncertain Conditions." *Cartography and Geographic Information Science* 36, no. 1 (2009): 29–43.

Roulin, N., and A. Bangerter. "Social Networking Websites in Personnel Selection." *Journal of Personnel Psychology* 12, no. 3 (2015): 143–51.

Roy, A. *Poverty Capital*. London: Taylor and Francis Group, 2012.

Rundstrom, R. "Teaching American Indian Geographies." In *Teaching American Ethnic Geography*, edited by L. Estaville and C. Rosen, 99–104. Washington DC: National Council for Geographic Education, 1997.

Ruppert, E. "The Governmental Topologies of Database Devices." *Theory, Culture and Society* 29 (2012): 116–36.

――――. "Rethinking Empirical Social Sciences." *Dialogues in Human Geography* 3, no. 3 (2013): 268–73.

Saleem, H., X. Yisha, and D. Ruths. "Effects of Disaster Characteristics on Twitter Event Signature." *Procedia Engineering* 78 (2014): 165–72.

Sandercock, L., and G. Attili. "Digital Ethnography as Planning Praxis: An Experiment with Film as Social Research, Community Engagement and Policy Dialogue." *Planning Theory and Practice* 11, no. 1 (2010): 23–45.

Sandvik, K., M. Jumbert, J. Karlsrud, and M. Kaufmann. "Humanitarian Technology: A Critical Research Agenda." *International Review of the Red Cross* 96, no. 693 (2014): 219–42. http://dx.doi.org/10.1017/s1816383114000344.

Sandvik, K., and K. Lohne. "The Rise of the Humanitarian Drone: Giving Content to an Emerging Concept." *Millennium: Journal of International Studies* (2014): 1–20.

Sayer, A. *Method in Social Science: A Realist Approach*. 2nd ed. London: Routledge, 1992.

Scharl, A., and K. Tochtermann. *The Geospatial Web: How Geobrowsers, Social Software and the Web 2.0 Are Shaping the Network Society.* London: Springer-Verlag, 2007.

Schlemper, M., J. Adams, and M. Solem. "Geographers in Business, Government, and Nonprofit Organizations: Skills, Challenges, and Professional Identities." *Professional Geographer* 66, no. 3 (2013): 480–92.

Schmidt, E., and J. Cohen. *The New Digital Age: Transforming Nations, Businesses, and Our Lives.* New York: Knopf, 2013.

Scholz, T. *Digital Labor: The Internet as Playground and Factory.* New York: Routledge, 2013.

Schonfeld, E. "Twitter Adds Places to Its Geo-Tweets, Just Don't Call It a Check-In." *TechCrunch*, April 14, 2010. http://techcrunch.com/2010/04/14/twitter-places -geo-tweets/.

Schroeder, R. "Big Data and the Brave New World of Social Media Research." *Big Data and Society* 1, no. 2 (2014). http://bds.sagepub.com/content/1/2 /2053951714563194.

Schuurman, N. "Database Ethnographies Using Social Science Methodologies to Enhance Data Analysis and Interpretation." *Geography Compass* 2, no. 5 (2008): 1529–48.

———. "Formalization Matters: Critical GIS and Ontology Research." *Annals of the Association of American Geographers* 96, no. 4 (2006): 726–39. http://www .jstor.org/stable/4124454.

———. *GIS: A Short Introduction.* Malden MA: Blackwell, 2004.

———. "Trouble in the Heartland: GIS and Its Critics in the 1990s." *Progress in Human Geography* 24, no. 4 (2000): 569–90.

———. "Work, Life, and Creativity among Academic Geographers." *Progress in Human Geography* (2009): 1–6. http://phg.sagepub.com/content/early/2009 /02/24/0309132508096350.full.pdf.

Schuurman, N., and G. Pratt. "Care of the Subject: Feminism and Critiques of GIS." *Gender, Place and Culture* 9, no. 3 (2002): 291–99.

Scott, J. *Seeing Like a State: How Certain Schemes to Improve the Human Condition Have Failed.* New Haven CT: Yale University Press, 1999.

Senaratne, H., L. Gerharz, E. Pebesma, and A. Schwering. "Usability of Spatio-Temporal Uncertainty Visualisation Methods." In *Bridging the Geographic Information Sciences*, edited by Jérôme Gensel, Didier Josselin, and Danny Vandenbroucke, 3–23. Berlin: Springer Berlin Heidelberg, 2012. http://link.springer.com.ezaccess .libraries.psu.edu/chapter/10.1007/978-3-642-29063-3_1.

Servon, L. *Bridging the Digital Divide: Technology, Community, and Public Policy.* Malden MA: Blackwell, 2002.

Severtson, D., and J. Myers. "The Influence of Uncertain Map Features on Risk Beliefs and Perceived Ambiguity for Maps of Modeled Cancer Risk from Air Pollution." *Risk Analysis* 33, no. 5 (2012): 818–37. doi:10.1111/j.1539-6924.2012.01893.x.

Shanley, L. "Opportunities and Challenges in Crisis Informatics." Paper presented to the White House Committee on Big Data, Washington DC, December 2012.

Shanley, L., R. Burns, Z. Bastian, and E. Robson. "Tweeting Up a Storm: The Promise and Perils of Crisis Mapping." *Photogrammetric Engineering and Remove Sensing* 79, no. 10 (2013): 865–79.

Shaw, A., S. Sheppard, S. Burch, D. Flanders, A. Wiek, J. Carmichael, J. Robinson, and S. Cohen. "Making Local Futures Tangible—Synthesizing, Downscaling, and Visualizing Climate Change Scenarios for Participatory Capacity Building." *Global Environmental Change* 19, no. 4 (2009): 447–63. doi:10.1016/j.gloenvcha.2009.04.002.

Shaw, J., and Mark Graham. "An Informational Right to the City? Code, Content, Control, and the Urbanization of Information." *Antipode*, first posted February 2, 2017. doi:10.1111/anti.12312.

Shekhar, S., S. Feiner, and W. Aref. "From GPS and Virtual Globes to Spatial Computing—2020." *GeoInformatica* (2013): 1–34.

Shekhar, S., V. Gunturi, M. Evans, and K. Yang. "Spatial Big-Data Challenges Intersecting Mobility and Cloud Computing." *MobiDE Eleventh ACM International Workshop on Data Engineering for Wireless and Mobile Access*, New York, 2013.

Shelly, K. "Social Media Makes It Easy to Share and Overshare Alike." *USA Today*, May 9, 2013.

Sheppard, E. "Knowledge Production through Critical GIS: Genealogy and Prospects." *Cartographica* 40, no. 4 (2005): 5–21.

———. "Quantitative Geography: Representations, Practices, and Possibilities." *Environment and Planning D: Society and Space* 19, no. 5 (2001): 535–54.

Sheppard, E., H. Couclelis, S. Graham, J. Harrington, and H. Onsrud. "Geographies of the Information Society." *International Journal of Geographical Information Science* 13, no. 8 (1999): 797–823.

Sheppard, S. "Landscape Visualisation and Climate Change: The Potential for Influencing Perceptions and Behaviour." *Environmental Science and Policy* 8, no. 6 (2005): 637–54.

Sheppard, S., A. Shaw, D. Flanders, S. Burch, A. Wiek, J. Carmichael, J. Robinson, and S. Cohen. "Future Visioning of Local Climate Change: A Framework for Community Engagement and Planning with Scenarios and Visualisation." *Futures* 43, no. 4 (2011): 400–412. doi:10.1016/j.futures.2011.01.009.

Shneiderman, B. "The Eyes Have It: A Task by Data Type Taxonomy for Information Visualizations." In *IEEE Symposium on Visual Languages, 1996: Proceedings*, 336–43. doi:10.1109/VL.1996.545307.

Short, J., R. Bohn, and C. Chaitanya. *How Much Information? 2010: Report on Enterprise Server Information*. San Diego: Global Information Industry Center, UC San Diego, January 2011. http://hmi.ucsd.edu/pdf/HMI_2010_EnterpriseReport_Jan_2011.pdf.

Shortridge, J. *Kansas City and How It Grew, 1822–2011*. Lawrence: University Press of Kansas, 2012.

Shrader, C. *History of Operations Research in the United States Army*. 3 vols. Washington DC: Office of the Deputy Undersecretary of the Army for Operations Research, United States Army, 2006. http://www.history.army.mil/catalog/browse/title.html#h.

Siegel, E. *Predictive Analytics*. Hoboken NJ: Wiley, 2013.

Silverman, B. *Density Estimation for Statistics and Data Analysis*. London: Chapman and Hall, 1986.

Simmel, G. "The Metropolis and Mental Life" (1903). In *On Individuality and Social Forms*, edited by Donald N. Levine, 324–39. Chicago: University of Chicago Press, 1971.

Simon, D. "How Small Data Will Revolutionize Communication." *Forbes*, March 6, 2014. http://www.forbes.com/sites/dansimon/2014/03/06/how-small-data-will-revolutionize-communications/.

Simon, P. 2013. *Too Big to Ignore: The Business Case for Big Data*. Hoboken NJ: John Wiley and Sons, 2013.

Singler, B. "'SEE MOM IT IS REAL': The UK Census, Jediism and Social Media." *Journal of Religion in Europe* 7, no. 2 (2014): 150–68. doi:10.1163/18748929-00702005.

Singleton, A., and S. Spielman. "The Past, Present and Future of Geodemographic Research in the United States and United Kingdom." *Professional Geographer* 66, no. 4 (2014): 558–67. doi:10.1080/00330124.2013.848764.

Skoric, M., N. Poor, P. Achanauparp, E. Lim, and J. Jiang. "Tweets and Votes: A Study of the 2011 Singapore General Election." *Proceedings of the 2012 45th Hawaii International Conference on System Sciences*, 2583–91. Washington DC: IEEE Computer Society, 2012.

Slocum, R. "Polar Bears and Energy-Efficient Lightbulbs: Strategies to Bring Climate Change Home." *Environment and Planning D* 22 (2004): 413–38.

Slovic, P., M. Finucane, E. Peters, and D. MacGregor. "Risk as Analysis and Risk as Feelings: Some Thoughts about Affect, Reason, Risk, and Rationality." *Risk Analysis* 24, no. 2 (2004): 311–22.

Smith, A. "46% of American Adults Are Smartphone Owners." *Pew Internet and American Life Project.* Washington DC: Pew Research Center, 2012. http://pewinternet .org/Reports/2012/Smartphone-Update-2012.aspx.

Smith, C. "Geospatial Encountering: Opportunistic Information Discovery in Web-Based GIS Environments." In *Proceedings of the American Society for Information Science and Technology* 48, no. 1 (2011): 1–4.

―――. "How Many People Use 900+ of the Top Social Media, Apps and Digital Services?" *Digital Marketing Stats/STrategy/Gadgets,* October 2015. http:// expandedramblings.com/index.php/resource-how-many-people-use-the-top -social-media/.

Smith, M., C. Szongott, B. Henne, and G. von Voigt. "Big Data Privacy Issues in Public Social Media." Paper presented at the 2012 6th IEEE International Conference on Digital Ecosystems and Technologies: Complex Environment Engineering, Campione d'Italia, June 18–20, 2012.

Smith, S. "Social Constructivisms and European Studies: A Reflectivist Critique." *Journal of European Public Policy* 6, no. 4 (1999): 682–91. doi:10.1080 /135017699343540.

Smolan, R., and J. Erwitt. *The Human Face of Big Data.* New York: Sterling, 2012.

Snickars, P. "If Content Is King, Context Is Its Crown." *VIEW Journal of European Television History and Culture* 1, no. 1 (2012): 1–6.

Solem, M., A. Kollasch, and J. Lee. "Career Goals, Pathways and Competencies of Geography Graduate Students in the USA." *Journal of Geography in Higher Education* 37, no. 1 (2013): 92–116.

Solove, D. "A Taxonomy of Privacy." *University of Pennsylvania Law Review* 154, no. 3 (2006): 477–560.

Sorte, F., and F. Thompson. "Poleward Shifts in Winter Ranges of North American Birds." *Ecology* 88, no. 7 (2007): 1803–12. doi:10.1890/06-1072.1.

Stalder, F. "Privacy Is Not the Antidote to Surveillance." *Surveillance and Society* 1, no.1 (2002): 120–24.

Standby Task Force. *Nepal Earthquake 2015* (blog). http://blog.standbytaskforce .com/deployments/nepal-earthquake-2015/.

Star, S., and K. Ruhleder. "Steps toward an Ecology of Infrastructure: Design and Access for Large Information Spaces." *Information Systems Research* 7, no. 1 (1996): 111–34.

Starbird, K., and L. Palen. "Working and Sustaining the Virtual 'Disaster Desk.'" Paper presented at the ACM 2013 Conference on Computer Supported Cooperative Work, San Antonio TX, February 23–27, 2013. https://cmci.colorado.edu/~palen/Home/Articles_by_Year_files/starbird-palen-cscw2013_final-2.pdf.

Stefanidis, A., A. Cotnoir, A. Crooks, M. Rice, and J. Radzikowski. "Demarcating New Boundaries: Mapping Virtual Polycentric Communities through Social Media Content." *Cartography and Geographic Information Science* 40, no. 2 (2013): 116–29.

Stefanidis, A., A. Crooks, and J. Radzikowski. "Harvesting Ambient Geospatial Information from Social Media Feeds." *GeoJournal* 78, no. 2 (2013): 319–38.

Stefik, M. *Internet Dreams: Archetypes, Myths, and Metaphors.* Cambridge MA: MIT Press, 1997.

Stephens, M. "Gender and the GeoWeb: Divisions in the Production of User-Generated Cartographic Information." *GeoJournal* 78, no. 6 (2013): 981–96.

Stocker, T., D. Qin, G. Plattner, M. Tignor, S. Allen, J. Boschung, A. Nauels, Y. Xia, V. Bex, and P. Midgley, eds. *Climate Change 2013: The Physical Science Basis.* Intergovernmental Panel on Climate Change, Working Group I Contribution to the IPCC Fifth Assessment Report (AR5). New York: Cambridge University Press, 2013. http://www.climatechange2013.org/images/report/WG1AR5_Frontmatter_FINAL.pdf.

Stottlemyre, S., and S. Stottlemyre. "Crisis Mapping Intelligence Information during the Libyan Civil War: An Exploratory Case Study." *Policy and Internet* 4, no. 3–4 (2012): 24–39.

Strauss, A., and J. Corbin. *Grounded Theory in Practice.* Thousand Oaks CA: SAGE, 1997.

Stroud, M. "The Minority Report: Chicago's New Police Computer Predicts Crimes, but Is It Racist?" *The Verge*, February 19, 2014. http://www.theverge.com/2014/2/19/5419854/the-minority-report-this-computer-predicts-crime-but-is-it-racist.

Sturgis, P., and N. Allum. "Science in Society: Re-Evaluating the Deficit Model of Public Attitudes." *Public Understanding of Science* 13, no. 1 (2004): 55.

Su, Y. S., C. Wardell, and Z. Thorkildsen. *Social Media in the Emergency Management Field: 2012 Survey Results.* Arlington VA: CNA Analysis and Solutions, 2013. http://www.cna.org/sites/default/files/research/SocialMedia_EmergencyManagement.pdf.

Sui, D. "Introduction: Strategic Directions for the Geographical Sciences in the Next Decade." *Professional Geographer* 63, no. 3 (2011): 305–9.

Sui, D., and R. Bednarz. "The Message Is the Medium: Geographic Education in the Age of the Internet." *Journal of Geography* 98, no. 3 (1999): 93–99.

Sui, D., and D. DeLyser. "Crossing the Qualitative-Quantitative Chasm I: Hybrid Geographies, the Spatial Turn, and Volunteered Geographic Information (VGI)." *Progress in Human Geography* 36, no. 1 (2012): 111–24.

Sui, D., S. Elwood, and M. Goodchild, eds. *Crowdsourcing Geographic Knowledge: Volunteered Geographic Information (VGI) in Theory and Practice.* London: Springer, 2012.

Sui, D., and M. Goodchild. "The Convergence of GIS and Social Media: Challenges for GIScience." *International Journal of Geographical Information Science* 25, no. 11 (2011): 1737–48.

Sundaresan, J., K. Santosh, A. Déri, R. Roggema, and R. Singh, eds. *Geospatial Technologies and Climate Change.* New York: Springer, 2004. http://www.springer.com/earth+sciences+and+geography/earth+system+sciences/book/978-3-319-01688-7.

Swim, J., S. Clayton, T. Doherty, R. Gifford, G. Howard, J. Reser, P. Stern, and E. Weber. "Psychology and Global Climate Change: Addressing a Multi-Faceted Phenomenon and Set of Challenges. A Report by the American Psychological Association's Task Force on the Interface between Psychology and Global Climate Change." Washington DC: American Psychological Association, 2009. http://www.apa.org/science/about/publications/climate-change.pdf.

Sylvester, D., and A. McGlynn. "The Digital Divide, Political Participation, and Place." *Social Science Computer Review* 28, no. 1 (2010): 64–74. doi:10.1177/0894439309335148.

Takacs, D. "How Does Your Positionality Bias Your Epistemology?" *NEA Higher Education Journal: Thought and Action* (Summer 2004): 27–38. http://ww.nea.org/assets/img/PubThoughtAndAction/TAA_03_04.pdf.

Tashakkori, A., and C. Teddlie, eds. *Handbook of Mixed Methodology in Social and Behavioral Research.* Thousand Oaks CA: SAGE, 2003.

Taylor, A. "The Square Kilometre Array." *Proceedings of the International Astronomical Union* 8, Symposium S291 (2012): 337–41.

Taylor, L., and R. Schroeder. "Is Bigger Better? The Emergence of Big Data as a Tool for International Development Policy." *GeoJournal* 80, no. 4 (2014): 503–18. http://link.springer.com/article/10.1007%2Fs10708-014-9603-5.

TechCrunch. "Twitter Cuts Off DataSift to Step Up Its Own Big Data Business." *TechCrunch*, April 11, 2015. http://techcrunch.com/2015/04/11/twitter-cuts-off-datasift-to-step-up-its-own-b2b-big-data-analytics-business/#.tewrlx:ufD7.

Tene, O., and J. Polonetsky. "Big Data for All: Privacy and User Control in the Age of Analytics." *Social Sciences Research Network*, 2012. http://ssrn.com/abstract=2149364.

Thakuriah, P., N. Tilahun, and M. Zellner. "Big Data and Urban Informatics: Innovations and Challenges to Urban Planning and Knowledge Discovery." In *Proceedings of NSF Workshop on Big Data and Urban Informatics* (2015): 4–32.

Thatcher, J. "Avoiding the Ghetto through Hope and Fear: An Analysis of Immanent Technology Using Ideal Types." *GeoJournal* 78, no. 6 (2013): 967–80.

———. "Big Data, Big Questions: Living on Fumes; Digital Footprints, Data Fumes, and the Limitations of Spatial Big Data." *International Journal of Communication* 8 (2014): 19. http://ijoc.org/index.php/ijoc/article/view/2174.

Thatcher, J., D. O'Sullivan, and D. Mahmoudi. "Data Colonialism through Accumulation by Dispossession: New Metaphors for Daily Data." *Environment and Planning D*, March 3, 2016. doi:0263775816633195.

Thoning, K., P. Tans, and W. Komhyr. "Atmospheric Carbon Dioxide at Mauna Loa Observatory: 2. Analysis of the NOAA GMCC Data, 1974–85." *Journal of Geophysical Research: Atmospheres* 94, no. D6 (1989): 8549–65. doi:10.1029/JD094iD06p08549.

Thrift, N. "The Place of Complexity." *Theory, Culture and Society* 16, no. 3 (1999): 31–69.

Thrift, N., and S. French. "The Automatic Production of Space." *Transactions of the Institute of British Geographers* 27, no. 3 (2002): 309–35. doi:10.1111/1475-5661.00057.

Thusoo, A., Z. Shao, S. Anthony, D. Borthakur, N. Jain, J. Sarma, R. Murthy, and H. Liu. "Data Warehousing and Analytics Infrastructure at Facebook." In *Proceedings of the 2010 ACM SIGMOD International Conference on Management of Data*, 1013–20. New York: Association for Computing Machinery (ACM), 2010.

Tokatli, N. "Old Firms, New Tricks, and the Quest for Profits: Burberry's Journey from Success to Failure and Back to Success Again." *Journal of Economic Geography* 12 (2012): 55–77.

Torrens, P. "Geography and Computational Social Science." *GeoJournal* 75, no. 2 (2010): 133–48.

Tsagkias, M., M. de Rijke, and W. Weerkamp. "News and Social Media." *WSDM '11 Proceedings of the Fourth ACM International Conference on Web Search and Data Mining*. New York: Association for Computing Machinery (ACM), 2011.

Tsou, M., D. Lusher, J. Yang, D. Gupta, J. Gawron, B. Spitzberg, L. An, and S. Wandersee. "Mapping Social Activities and Concepts with Social Media (Twitter) and Web Search Engines (Yahoo and Bing): A Case Study in 2012 US Presidential Election." *Cartography and Geographic Information Science* 40, no. 4 (2013): 337–48.

Tufts, S., W. Jacobson, and M. Stevens. "Status Update Social Media and Local Government Human Resource Practices." *Review of Public Personnel Administration* 35, no. 2 (2015): 193–207.

Tumasjan, A., T. Sprenger, P. Sandner, and I. Welpe. "Predicting Elections with Twitter: What 140 Characters Reveal about Political Sentiment." Paper presented at the Fourth International AAAI Conference, Washington DC, May 23–26, 2010.

Turner, F. *From Counterculture to Cyberculture: Stewart Brand, the Whole Earth Network, and the Rise of Digital Utopianism.* Chicago: University of Chicago Press, 2008.

UK Census. *Irish in Britain: How Irish Are You?* UK Census, Office of National Statistics, 2011. http://www.irishinbritain.org/campaigns/how-irish-are-you-census-2011/.

United Nations. Office for the Coordination of Humanitarian Affairs (UN OCHA). *Humanitarianism in the Network Age.* Geneva: UN Office for the Coordination of Humanitarian Affairs, 2013. https://ochanet.unocha.org/p/Documents/WEB Humanitarianism in the Network Age vF single.pdf.

U.S. Census Bureau. *2010 U.S. Census.* http://www.census.gov/2010census/.

———. *American Community Survey.* Washington DC: U.S. Census Bureau, 2012.

Valentine, G. "A Corporeal Geography of Consumption." *Environment and Planning D: Society and Space* 17 (1999): 329–51.

Valiela, I., and J. Bowen. "Shifts in Winter Distribution in Birds: Effects of Global Warming and Local Habitat Change." *AMBIO: A Journal of the Human Environment* 32, no. 7 (2003): 476–80. doi:10.1579/0044-7447-32.7.476.

Van Gorp, A. "Integration of Volunteer and Technical Communities into the Humanitarian Aid Sector: Barriers to Collaboration." In *Proceedings of the 11th International ISCRAM Conference,* edited by S. R. Hiltz, M. S. Pfaff, L. Plotnick, and P. C. Shih, 622–31. University Park: Pennsylvania State University, 2014.

Verity, A. *OCHA's Lessons Learned: Collaboration with V&TCs for Libya and Japan.* Geneva: United Nations Office for the Coordination of Humanitarian Affairs, 2011. http://digitalhumanitarians.com/sites/default/files/resource-field_media/OCHALessonsLearnedCollaborationwithVTCsinLibyaandJapanFinalNov2011.pdf.

Viard, T., G. Caumon, and B. Lévy. "Adjacent versus Coincident Representations of Geospatial Uncertainty: Which Promote Better Decisions?" *Computers and Geosciences* 37, no. 4 (2011): 511–20.

Vieweg, S. "Situational Awareness in Mass Emergency: A Behavioral and Linguistic Analysis of Microblogged Communications." PhD diss., University of Colorado,

2012. http://works.bepress.com/cgi/viewcontent.cgi?article=1029andcontext =vieweg.

Vieweg, S., A. Hughes, K. Starbird, and L. Palen. "Microblogging during Two Natural Hazards Events: What Twitter May Contribute to Situational Awareness." Paper presented at CHI 2010: Crisis Informatics, Atlanta, April 13, 2010.

Virtual Social Media Working Group, and DHS First Responders Group. *Lessons Learned: Social Media and Hurricane Sandy*. U.S. Department of Homeland Security, 2013. http://www.ghinternational.com/docs/DHS_VSMWG_Lessons _Learned_Social_Media_and_Hurricane_Sandy_Formatted_June_2013 _FINAL.pdf.

Volkova, S., B. van Durme, D. Yarowsky, and Y. Bachrach. "Social Media Predictive Analytics." In *Proceedings of the Conference of the North American Chapter of the Association for Computational Linguistics—Human Language Technologies*. Denver: Association of Computational Linguistics, 2015.

Waldman, A., A. Verity, and S. Roberts. *Guidance for Collaborating with Formal Humanitarian Organizations*. Geneva: Digital Humanitarian Network, 2013. https://app .box.com/s/w25sqotkg4qc2f2ch1ii.

Waldrop, M. *Complexity: The Emerging Science at the Edge of Order and Chaos*. New York: Simon and Schuster, 1992.

———. "Small Data Is Beautiful." *DataScienceCentral*, April 10, 2013. http://www .datasciencecentral.com/profiles/blogs/small-data-is-beautiful.

Wallace, D., and R. Wallace. *A Plague on Your Houses: How New York Was Burned Down and National Public Health Crumbled*. London: Verso, 2011.

Waller, V. "Information Systems 'In the Wild': Supporting Activity in the World." *Behaviour and Information Technology* 28, no. 6 (2009): 577–88. doi:10.1080 /01449290903188971.

Wallis, J., E. Rolando, and C. Borgman. "If We Share Data, Will Anyone Use Them? Data Sharing and Reuse in the Long Tail of Science and Technology." *PLOS ONE* 8, no. 7 (2013): e67332.

Wang, S., L. Anselin, B. Badhuri, C. Crosby, M. Goodchild, Y. Liu, and T. Nyerges. "CyberGIS Software: A Synthetic Review and Integration Roadmap." *International Journal of Geographical Information Science* 27, no. 11 (2013): 2122–45.

Wang, S., and M. Armstrong. "A Theoretical Approach to the Use of Cyberinfrastructure in Geographical Analysis." *International Journal of Geographical Information Science* 23, no. 2 (2009): 169–93. doi:10.1080/13658810801918509.

Warf, B. "Contemporary Digital Divides in the United States." *Tijdschrift voor Econ-omische en Sociale Geografie* 104, no. 1 (2013): 1–17. doi:10.1111/j.1467-9663.2012 .00720.x.

———. "Googlebombs, Warblogs, and Hacktivism: The Internet as Agent for Progressive Social Change." In *Nonkilling Geographies*, edited by J. Tyner and J. Inwood, 127–48. Honolulu: Center for Global Nonkilling, 2011.

Waring, M. *If Women Counted: A New Feminist Economics*. San Francisco: Harper and Row, 1988.

Weber, E. "Experience-Based and Description-Based Perceptions of Long-Term Risk: Why Global Warming Does Not Scare Us (Yet)." *Climatic Change* 77, no. 1 (2006): 103–20.

Weidemann, C., and J. Swift. "Social Media Location Intelligence: The Next Privacy Battle—An ArcGIS Add-in and Analysis of Geospatial Data Collected from Twitter .com." *International Journal of Geoinformatics* 9, no. 2 (2013): 21–27.

Weiler, A., M. Grossniklaus, and M. Scholl. "Situation Monitoring of Urban Areas Using Social Media Data Streams." *Information Systems* (forthcoming).

Weinberger, D. *Too Big to Know: Rethinking Knowledge Now That the Facts Aren't the Facts, Experts Are Everywhere, and the Smartest Person in the Room Is the Room.* New York: Basic Books, 2011.

Weisberg, M. *Simulation and Similarity: Using Models to Understand the World*. Oxford: Oxford University Press, 2013.

Weiss, M. J. *The Clustering of America*. New York: Harper and Row, 1988.

Weiss, S., and N. Indurkhya. *Predictive Data Mining: A Practical Guide*. San Francisco: Morgan Kaufmann, 1998.

Wellen, C., and R. Sieber. "Toward an Inclusive Semantic Interoperability: The Case of Cree Hydrographic Features." *International Journal of Geographical Information Science* 27, no. 1 (2013): 168–91.

Wilken, R. "Mobilizing Place: Mobile Media, Peripatetics, and the Renegotiation of Urban Places." *Journal of Urban Technology* 15, no. 3 (2008): 39–55. doi:10.1080 /10630730802677939.

Williams, A., and P. Dourish. "Imagining the City: The Cultural Dimensions of Urban Computing." *Computer* 39, no. 9 (2006): 38–43. doi:10.1109/mc.2006.313.

Willis, K., C. Holscher, G. Wilbertz, and C. Li. "A Comparison of Spatial Knowledge Acquisition with Maps and Mobile Maps." *Computers, Environment and Urban Systems* 33, no. 2 (2009): 100–110.

Wilson, M. "Data Matter(s): Legitimacy, Coding, and Qualifications-of-Life." *Environment and Planning D: Society and Space* 29, no. 5 (2011): 857–72.

———. "Location-Based Services, Conspicuous Mobility, and the Location-Aware Future." *Geoforum* 43, no. 6 (2012): 1266–75. http://dx.doi.org/10.1016/j.geoforum .2012.03.014.

———. "Morgan Freeman Is Dead and Other Big Data Stories." *Cultural Geographies* 22, no. 2 (2015): 345–49.

———. "Towards a Genealogy of Qualitative GIS." In *Qualitative GIS: A Mixed Methods Approach*, edited by M. Cope and S. Elwood, 156–70. London: SAGE, 2009.

———. "'Training the Eye:' Formation of the Geocoding Subject." *Social and Cultural Geography* 12 (2011): 357–76.

Winner, L. *The Whale and the Reactor: A Search for Limits in an Age of High Technology.* Chicago: University of Chicago Press, 1986.

Winsberg, E. *Science in the Age of Computer Simulation.* Chicago: University of Chicago Press, 2010.

Woodrow Wilson Center. *Legal and Policy Issues* (video). Washington DC, October 24, 2012. http://youtu.be/apEMNJFnBEM.

———. *Research Challenges* (video). Washington DC, November 2, 2012. http://youtu.be/uTIT3mkQhew.

Wright, D., M. Goodchild, and J. Proctor. "Demystifying the Persistent Ambiguity of GIS as 'Tool' versus 'Science.'" *Annals of the Association of American Geographers* 87, no. 2 (1997): 346–62.

Wu, M. "The Big Data Fallacy and Why We Need to Collect Even Bigger Data." *TechCrunch*, November 25, 2012. http://techcrunch.com/2012/11/25/the-big -data-fallacy-data-%e2%89%a0-information-%e2%89%a0-insights/.

Wyly, E. "Automated (Post)Positivism." *Urban Geography* 35, no. 5 (2014): 669–90.

———. "The New Quantitative Revolution." *Dialogues in Human Geography* 4, no. 1 (2014): 26–38.

———. "Strategic Positivism." *Professional Geographer* 61, no. 3 (2009): 310–22.

Xiao, Y., Q. Huang, and K. Wu. "Understanding Social Media Data for Disaster Management." *Natural Hazards* 79, no. 3 (2015): 1–17.

Xu, C., D. Wong, and C. Yang. "Evaluating the 'Geographical Awareness' of Individuals: An Exploratory Analysis of Twitter Data." *Cartography and Geographic Information Science* 40, no. 2 (2013): 103–15.

Yakabuski, K. "Big Data Should Inspire Humility, Not Hype." *Globe and Mail*, March 4, 2013, A11.

Young, J., and M. Gilmore. "Subaltern Empowerment in the Geoweb: Tensions between Publicity and Privacy." *Antipode* 46, no. 2 (2014): 574–91.

Young, J., D. Wald, P. Earle, and L. Shanley. *Transforming Earthquake Detection and Science through Crowdsourcing.* Washington DC: Woodrow Wilson International

Center for Scholars, 2013. https://www.wilsoncenter.org/sites/default/files
/CitizenSeismology_FINAL.pdf.

Zheng, Liu, and T. Hsieh. "U-Air: When Urban Air Quality Inference Meets Big Data." Microsoft Research, 2014. http://research.microsoft.com/apps/pubs /default.aspx?id=193973.

Zhong, C., M. Batty, E. Manley, J. Wang, Z. Wang, F. Chen, and G. Schmitt. "Variability in Regularity: Mining Temporal Mobility Patterns in London, Singapore and Beijing Using Smart-Card Data." *PLOS One*, February 12, 2016. http://journals .plos.org/plosone/article?id=10.1371/journal.pone.0149222.

Zickuhr, K. "Three-Quarters of Smartphone Owners Use Location Based Services." *Pew Internet and American Life Project*. Washington DC: Pew Research Center, 2012. http://www.pewinternet.org/files/old-media/Files/Reports/2012/PIP _Location_based_services_2012_Report.pdf.

Ziemke, J. "Crisis Mapping: The Construction of a New Interdisciplinary Field?" *Journal of Map and Geography Libraries: Advances in Geospatial Information, Collections and Archives* 8, no. 2 (2012): 101–17.

Zikopoulos, P., C. Eaton, D. deRoos, T. Deutsch, and G. Lapis. *Understanding Big Data: Analytics for Enterprise Class Hadoop and Streaming Data*. New York: McGraw-Hill, 2012.

Zook, M., and M. Graham. "The Creative Reconstruction of the Internet: Google and the Privatization of Cyberspace and DigiPlace." *Geoforum* 38 (2007): 1322–43.

―――. "Mapping DigiPlace: Geocoded Internet Data and the Representation of Place." *Environment and Planning B: Planning and Design* 34 (2007): 466–82.

Zook, M., M. Graham, T. Shelton, and S. Gorman. "Volunteered Geographic Information and Crowdsourcing Disaster Relief: A Case Study of the Haitian Earthquake." *World Medical and Health Policy* 2, no. 2 (2012): 7–33.

Zuckerman, E. "The Internet's Original Sin." *The Atlantic*, August 2014. http://www .theatlantic.com/technology/archive/2014/08/advertising-is-the-internets -original-sin/376041/4/.

Zuk, T., and S. Carpendale. "Theoretical Analysis of Uncertainty Visualizations." *Proceedings of SPIE* 6060 (2006): 66–79.

CONTRIBUTORS

Ryan Burns is assistant professor of geography at the University of Calgary, with research and teaching interests in social and political intersections of spatial technology. He has recently explored the inequalities and uneven impacts of digital humanitarianism, big data, open data, and smart cities. He has published articles in *Geoforum, GeoJournal, Environment & Planning A,* and *Photogrammetric Engineering & Remote Sensing;* as well, he has co-organized special issues of *GeoJournal,* ACME: *An International e-Journal for Critical Geographies, Computational Culture,* and *Professional Geographer.* He has also authored policy-related papers for the Woodrow Wilson International Center for Scholars and the UN *Human Development Report.* Before coming to Calgary, he worked as assistant professor of instruction at Temple University, earned his PhD from the University of Washington, and received his MS from San Diego State University. He has served as the secretary/treasurer and student representative for the Geographic Information Systems & Science specialty group and as webmaster of the Socialist & Critical Geography specialty group of the American Association of Geographers.

Josef Eckert is a PhD candidate in the Department of Geography at the University of Washington. His research interests include online social activism, qualitative analysis of big data derived from social media, and, increasingly, the ways in which students use technology to developmentally inform their educational processes.

Emily Fekete is a clinical assistant professor in the Department of Geography at Oklahoma State University. Her research interests largely focus on Internet geographies and the economy. Related articles include "Race and (Online) Sites of Consumption," published in the *Geographical Review* (2015), and "Consumption and the Urban Hierarchy in the Southeastern United States," published in *Southeastern Geographer* (2014). Her other publications discuss the use of Internet technologies for political protest, as well as cyberwarfare and cyberterrorism, with a focus on current events in the Middle East.

Mark Graham is professor of Internet geography at the Oxford Internet Institute, a faculty fellow at the Alan Turing Institute, a research fellow at Green Templeton College, an associate in the University of Oxford's School of Geography and the Environment, and a visiting fellow at the Department of Media and Communications in the London School of Economics and Political Science. He leads a range of research projects spanning topics such as digital labor, the gig economy, Internet geographies, and ICTs and development. His research is described at www.markgraham.space.

Jin-Kyu Jung is an associate professor in the School of Interdisciplinary Arts and Sciences at the University of Washington Bothell and an adjunct associate professor in the Department of Geography at the University of Washington. He is an urban geographer whose interdisciplinary research program contributes to qualitative geographic information systems (GIS)/geovisualization, children's urban geographies, and community-based research and planning. He has particularly focused on developing new ways of expanding these critical, qualitative, and creative capabilities of GIS and geographic visualization in conjunction with the social and spatial theorization of (urban) space. He received a PhD in geography and a master's in urban planning from State University of New York at Buffalo (University at Buffalo SUNY) and a BA in urban engineering from Pusan National University in South Korea.

Matthew Kelley is an associate professor of urban studies and geographic information systems (GIS) at the University of Washington Tacoma. He received his PhD in geography from Pennsylvania State University and in his research and teaching has focused on the impacts and applications of geospatial technologies in urban space. In particular his research focuses on (1) relational socio-cultural processes that are implicated in the production of urban space, (2) participatory approaches to deploying GIS in urban and community development scenarios, and (3) emergent issues of digital exclusion in urban areas (new digital divides). At UW Tacoma he directs an undergraduate certificate program in GIS and a graduate program in geospatial technologies.

Karen Kemp is professor of the practice of spatial sciences at the University of Southern California. From her home base in Hawaii, she currently teaches full-time in the online MS GIS program. While her teaching at USC is focused largely on spatial analysis and spatial modeling, her research and publishing areas include geohumanities, environmental modeling, and GIS education. Some of her major contributions include coediting the influential 1990 NCGIA Core Curriculum in

288   Contributors

GIS, serving on the editorial team for the UCGIS *Geographic Information Science and Technology Body of Knowledge* (2006), and editing SAGE's *Encyclopedia of Geographic Information Science* (2008).

Rob Kitchin is professor and ERC advanced investigator at the National University of Ireland, Maynooth. He is (co)principal investigator of the Programmable City project, the Building City Dashboards project, the All-Island Research Observatory, and the Digital Repository of Ireland. He has published widely across the social sciences, including 23 authored/edited books and more than 150 articles and book chapters. He was the editor-in-chief of the 12-volume *International Encyclopedia of Human Geography* and is currently editor of the journal *Dialogues in Human Geography*. He was the 2013 recipient of the Royal Irish Academy's Gold Medal for the Social Sciences.

Tracey Lauriault is assistant professor of critical media and big data, media studies, and communication in the School of Journalism and Communication at Carleton University. Her areas of expertise are critical data studies, small, big, and spatial data policy, data infrastructures and open data, open government, geospatial data, crowd-sourcing, and the preservation and archiving of data. She is a research associate with the European Research Council–funded Programmable City project in the area of political arithmetic, territorial geometry, and programmed cities, which entails three case studies aiming to determine how digital data about cities and their citizens are materially and discursively supported and processed. She is also a research associate with the Geomatics and Cartographic Research Centre at Carleton University. She is a steering committee member of Research Data Canada, on the board of Open North, and a member of the Institute for Data Science at Carleton. She received the 2016 Inaugural Open Data Leadership Award for Canada.

Jessa Lingel is an assistant professor at the Annenberg School for Communication and Journalism at the University of Pennsylvania. She received her PhD in communication and information from Rutgers University. She has an MLIS from Pratt Institute and an MA from New York University. Her research interests include information inequalities and technological distributions of power.

David O'Sullivan is associate professor of geography and Chancellor's Professor at the University of California, Berkeley. His research interests are in simulation models and geographic complexity, urban neighborhood change, and the social and political implications of geospatial technology. He is the author of more than fifty peer-reviewed papers and book chapters, and coauthor (with David Unwin) of *Geographic Information Analysis* (2010) and (with George Perry) of *Spatial Simulation: Exploring Pattern and Process* (2013).

David Retchless is an assistant professor of marine sciences at Texas A&M University at Galveston. Trained as a geographer at Penn State University, he is interested in the cartographic communication of climate change and coastal hazards, with a particular focus on storm surge and sea-level rise. He takes a user-centered approach to evaluating traditional and web-based cartographic products, deploying survey, focus group, and think-aloud methods to better understand how these products interact with individual differences to (re)shape perceptions of coastal vulnerability and risk.

Britta Ricker is an assistant professor in the urban studies program at the University of Washington Tacoma. She earned her PhD in geography at Simon Fraser University in 2014. She earned a master of science in geography at McGill University and a bachelor of science with a dual major in geography and international politics from Frostburg State University in Maryland. She has taught a wide variety of courses related to geographic information systems (GIS) and science. Her research interests converge around spatial information collection and dissemination opportunities afforded by mobile computers. She is interested in applying them to spatial learning related to emergency preparedness and environmental communication initiatives. Her professional experience includes acting as a hazard mapping analyst for Dewberry and Davis, a firm that serves as consultant for the Federal Emergency Management Agency (FEMA). In this role she provided GIS services and collected data in the Gulf Coast region after Hurricanes Katrina and Rita in 2005 and 2006. She has also acted as a cartographic consultant for the Commission for Environmental Cooperation (CEC) and provided cartographic services (on paper!) for MapQuest.

Andy Shears is an assistant professor of geography in the Department of Geosciences at Mansfield University. His interests in geography are wide ranging but tend to fall at the intersection of technology, mapping, media, power, critical geopolitics, and autoethnography. His work has explored the representation of geographical imagination in American television, the impact of fatness on conceptions of place, and the role of location-based social media on personal autobiographical narratives. Shears is currently transitioning out of the higher education industry to seek more challenging intellectual opportunities; to that end, he is the founder and principal of Muncie Map Company, a geospatial consulting and cartographic research firm based in his hometown of Muncie, Indiana. In his spare time Shears enjoys making maps of bizarre topics, playing the ukulele, hacking vintage Apple II computers for modern purposes, and rehabilitating and fostering homeless dogs for adoption.

Jungyeop Shin is professor in and chair of the Department of Geography Education and an adjunct professor in the Interdisciplinary Program of Environmental Education at Seoul National University, Seoul, South Korea. He received his PhD from the Department of Geography at the State University of New York at Buffalo in 2005. His research interests are particularly centered on polynucleated urban structure, GIS mapping, spatial analysis, and spatial cognition.

Renee E. Sieber is associate professor, jointly appointed between the Departments of Geography and Environment at McGill University in Montreal. Her research is situated at the intersection of social theory and software architecture. In 2016 she was given the lifetime achievement award and GIScience excellence award from the Canadian Association of Geographers. She wrote the definitive framing of public participation geographic information systems ("Public Participation Geographic Information Systems: A Literature Review and Framework"), which has been cited more than seven hundred times. Her current research involves the ways in which open data, crowdsourcing, and the geospatial web are shaping the relationships between cities and citizens.

Jennifer Swift is associate professor of teaching of spatial sciences at the University of Southern California. Residing in Los Angeles, she teaches full-time in the undergraduate residential bachelor of science in geodesign and online GIS programs in the Spatial Sciences Institute (SSI). As founding director of graduate studies for the geographic information science and technology (GIST) programs, she developed key systems and protocols to ensure that staff and faculty utilize high-quality teaching practices and seek out and share innovative research opportunities for all SSI students. She has coedited conference proceedings and technical reports on data modeling, archiving, and dissemination of geotechnical data, including *PEER Lifelines Geotechnical Virtual Data Center, Pacific Earthquake Engineering Research Center Report (Berkeley) PEER 108*, and other publications in earthquake research. Her teaching, research, and publications are currently focused on teaching web GIS and cyberGIS and improving methods for teaching programming across disciplines.

Matthew Tenney is a PhD candidate in the Department of Geography at McGill University. His research on "coded engagement" takes a broad look at how society and technology are converging with transformative impacts on all aspects of everyday life, as well as how these forces are redefining the practice and study of geography more generally. Over the past few years he has designed technical frameworks for interrogating aspects of connectivity between citizens, communities, and city government through spatial and data-driven networks. His work has aimed to develop analytical tools for data useful in measuring social insights, stakeholder engagement, and social capital that aim to solve real-world problems.

Jim E. Thatcher is an assistant professor of urban studies at the University of Washington Tacoma. His research examines the recursive relations among extremely large geospatial data sets, the creation and analysis of those data sets, and society, with a focus on how data have come to mediate, saturate, and sustain modern urban environments. His work takes as central questions how the constitution and analysis of data are framed by and subsequently frame the production of knowledge with regard to nature, society, and politics. Often referred to as critical data studies or digital political ecologies, his work has been featured in national media outlets, including NPR and *The Atlantic*. He is profoundly mediocre at chess and terrible at banjo but enjoys both. This is his first book.

Chris Weidemann is the director of information technology at HGL, Inc. As a federal contractor located in the Washington DC metropolitan area, he provides information technology and geospatial support to multiple branches of the Department of Defense, the Environmental Protection Agency, the Department of Justice, and the Department of Energy. He graduated from the University of Southern California's Geographic Information Science and Technology master's program. He previously published an article on social media location intelligence and privacy in the *International Journal of Geoinformatics* in June 2013. Some of his ongoing research topics include open source intelligence gathering, natural language processing, and big data analytics.

INDEX

accumulation by dispossession, xv, xxiii

algorithms, xiii, xiv, 18, 24, 170, 175; bias, 62; contrasts with human mind, 49, 56, 188; monitoring of, xv, 235; uses of, 28, 118, 141, 174

ambient geospatial information, 126–27

anonymity, 170

anticipatory governance, 16

application programming interface (API), 149, 219

Arar, Maher, 14

Benn, Tony, 235

big data, 16, 21, 34, 41, 72, 91, 116–17, 123, 145, 165, 198, 214; boosters of, xi; definitions of, xii, xiii, 4, 23, 28, 43, 231; epistemologies of, xiii–xiv, xvii, 23–24, 34, 47, 214, 221; ethics of, 170, 175, 206; histories of, xiv, 23, 42–43; infrastructures, xxi, 73; limits of, xviii; normative dimensions of, 173; paradigms, 174; three Vs of, xiii, 23, 44, 65, 72. *See also* data; small data

black box, 71, 80, 83–84

capital accumulation, xi, xvi, 7, 25, 35

capitalist modernity, xvii, xxiii

citizen science, 199, 204–6

climate change, xxi, 197–99, 209; communication, 199, 209; geography, 199;

models, 198, 202; public views of, 204; science, 208; vulnerability, 210

complexity science, 22, 30, 33

complexity theory, 29

consent, xv; in research, 176, 206

consumption, 15, 145, 150, 221

context matching, 133

critical cartography, 79, 204

critical data studies, xii, xviii, 7, 17–19, 215

critical GIS, 77

cyborg, 80

data, 3, 33; accuracy, 216; analytics, 41; assemblages, 8–13, 18; brokers, 5; commodity, 7, 147; deluge, xi, 21, 23, 43, 49; demographic, 164; fumes, 74; geospatial, 197, 198; infrastructures, 5, 214; interoperability, 59; materiality, 170, 231; mining, 104–6, 124, 198; ontology, 11; open, 5, 7, 18; political economies of, 8, 10, 215; provenance, 74; raw, xiv, 47, 63, 83, 235; rights, 50; science, 92; shadows, 15; sharing, 59; spatial, 29; topology, 48; vis-à-vis information, 55. *See also* big data; geodemographics; ontology; small data

databases, 3, 6, 75, 175, 224

dataveillance, 15

digital divide, 165, 178–79, 188; definition, xxi, 178

Jedi Knight, 57

Kansas City, 152–53
kernel density, 106, 108
knowledge politics, 220, 223–24
knowledge production, xvii, 3, 33, 41, 70, 215

location-based services (LBS), 73, 93, 128, 184
locative media, 5, 172
looping effect, 11, 14

machine learning, 7, 15, 25, 62, 118, 124, 138
"male gaze," 147
marginalization, 190
marijuana, 101
materiality, 170. *See also* data: materiality
metadata, 51, 58–59, 82, 93, 136
methods, 21, 42, 126, 136; participatory, 190, 219; qualitative, xvii, 18, 60, 71, 76, 83, 92–93; quantitative, xvii, 22. 25–27, 36, 76, 92, 95, 215; mixed, xvii, 81, 93, 96–99, 116–17, 119, 232; reflexive, 71
metronormativity, 169
Metropolitan Transit Authority (MTA), 171–72
model-based approaches, 32, 234

natural disasters, 94, 140–41, 148, 208, 218
neogeography, xi
New Zealand, 23

objectivity, 70, 78 , 83
ontology, xiii, 55, 97
open data. *See* big data, data
Openshaw, Stan, 26
OpenStreetMap (OSM), 140, 147, 199, 214

Palin, Sarah, 126
Pearson correlation, 151
performativity, 224
play, 170
positionality, 64, 70–71, 78, 82. *See also* subjectivity
positivism, xvii, 24, 71, 76, 78, 83, 97
power/knowledge, 6, 9
pragmatism, 119
privacy, 123, 142, 169, 170, 189, 206, 209, 233; laws, 174
processes, 22, 32
public health, 94

qualitative GIS, 77, 97
quantitative revolution, 29

reflexivity, 70, 80
remote sensing, 200–201

Santa Fe Institute (SFI), 30
scale, xix, xxi, 13, 24, 32, 61, 190, 209; human, 199; spatial, 221; spatio-temporal, 200; temporal, 95
Silicon Valley, 30
simulation, 28–29
situated knowledges, 78, 91
Sleater Kinney, 172
small data, xv, 29, 198, 232; definition of, 4, 45, 47–48, 52–54, 199; histories of, 45–46
smart cities, 23
Snowden, Edward, 14
social constructivism, 70
social justice, 175, 210
social media, xvii, xx, 16, 51, 56, 118, 123; predictive uses of, 94, 125, 145; privacy issues, 124–25, 139–42, 169; research questions, 206; spatial, 93,

www.ingramcontent.com/pod-product-compliance
Lightning Source LLC
Chambersburg PA
CBHW020337270326
41926CB00007B/211